Zinc-Air Batteries

Shengjie Peng

Zinc-Air Batteries

Fundamentals, Key Materials and Application

 Springer

Shengjie Peng
Department of Applied Chemistry
Nanjing University of Aeronautics
and Astronautics
Nanjing, China

ISBN 978-981-19-8216-3   ISBN 978-981-19-8214-9   (eBook)
https://doi.org/10.1007/978-981-19-8214-9

This Springer imprint is published by the registered company Springer Nature Singapore Pte Ltd.
The registered company address is: 152 Beach Road, #21-01/04 Gateway East, Singapore 189721,
Singapore

Preface

Over the last few decades, electrochemical energy storage materials and devices have developed at an unprecedented rate, driven by consumer electronics, power tools and, more recently, automotive and renewable energy storage. Metal-air batteries represent one of the most efficient energy storage technologies due to outstanding features such as high round-trip efficiency, long life cycle, fast response during peak electricity demand/supply periods and reduced weight using atmospheric oxygen as one of the main reactants. Depending on the metal anode, metal-air batteries can be divided into several types, such as zinc-air batteries, magnesium-air batteries, aluminum-air batteries and lithium-air batteries. Zinc-air batteries are the preferred choice among all metal-air batteries for their low cost, lightweight, large scale, high energy density, safer battery technology and environmental friendliness. Rechargeable zinc-air batteries are regarded as very important rising energy storage systems due to their utility in portable electronics, grid management and electric vehicles.

This book titled *Zinc-Air Batteries: Fundamentals, Key Materials and Application* aims to discuss the cutting-edge materials and technologies for zinc-air batteries. From the perspective of basic research and engineering application, the principle innovation, research progress and technical breakthrough of key materials such as positive and negative electrodes, electrolytes and separators of zinc-air batteries are discussed systematically. This book also involves the theoretical calculation methods and advanced characterization techniques, which provides theoretical and technical references for the design and development and performance improvement of different zinc-air battery systems based on materials optimization. Current technologies, achievements and challenges and future development directions are further discussed, which can be used to guide and promote the development of zinc-air battery technology. This book is suitable to researchers at all levels in the energy area and provides them with a quick way of understanding the development of zinc-air batteries. Finally, we express our sincere thanks to all the contributors to this book, who are actively engaged in cutting-edge research on metal-air batteries.

Nanjing, China   Shengjie Peng

Contents

About the Author

Prof. Shengjie Peng (FRSC) received his Ph.D. degree at Nankai University (P. R. China) in 2010. Following a postdoctoral fellowship with Prof. Alex Yan and Prof. Seeram Ramakrishna in Nanyang Technological University and National University of Singapore, he is now working as a professor at Nanjing University of Aeronautics and Astronautics. His current research interests focus on development of rationally designed functional materials with finely tailored nanoscale architecture to tackle critical problems (such as energy density, power density, cycle and calendar life, safety and cost) in diverse energy-related applications, including ORR, water splitting, batteries, fuel cells, as well as clean and renewable energy. To date, he has co-authored 160 peer-reviewed publications with over 10,000 citations, 55 H-index.

Chapter 1
Overview of Zinc-Air Battery

1.1 History of Zinc-Air Battery

Energy is the material basis for the progress and development of human civilization. Since the industrial revolution, with the gradual consumption of fossil energy and the increasingly prominent environmental pollution problem, the demand for green, clean and renewable energy has grown rapidly, and the energy system has shown a trend of transformation from the absolute dominance of fossil energy to low-carbon and multi-energy integration [1, 2]. In recent years, electrochemical energy storage technology that has maintained a rapid growth momentum is considered as a sustainable and environmentally friendly green energy, and can be well used as a medium for energy storage and conversion, so it has aroused widespread concern in emerging markets and scientific research fields. This electrochemical energy storage technology is not limited by geographical environment, and can directly store and release electrical energy, mainly including various batteries, lead-acid batteries, lithium-ion batteries, sodium-sulfur batteries, flow batteries, etc. The market scale of these batteries is increasingly mature, and the application field is expanding, suitable for a wide range of occasions. Among them, metal-air batteries, as an emerging electrochemical energy storage technology, are expected to become a new generation of energy storage devices due to their abundant raw materials, low cost, high efficiency, and pollution-free characteristics.

Metal-air batteries use metals such as magnesium, aluminum, zinc, mercury, iron as the negative electrode, oxygen in the air or pure oxygen as the positive active substance. They play an important role in today's national economy and are widely used in industry, agriculture, transportation, post and telecommunications, etc. Among various metal anodes, zinc electrodes are considered to have very broad development prospects due to their abundant reserves, low price, good corrosion resistance, and ideal reaction kinetics in alkaline aqueous solutions. In 1879, Maiche et al. assembled the world's first zinc-air battery using metal zinc as the anode, platinum-plated carbon electrode as anode on the contrary of the manganese dioxide in Le Kronschie battery, in a slightly acidic aqueous solution of NH_4Cl as electrolyte.

S. Peng, *Zinc-Air Batteries*, https://doi.org/10.1007/978-981-19-8214-9_1

The structure and principle are as follows:

$$Zn|NH_4Cl|O_2(C)$$
$$Zn + 2NH_4Cl + \tfrac{1}{2}O_2 \rightarrow Zn(NH_3)_2Cl_2 + H_2O$$

The structure and appearance of this zinc-air battery are similar to zinc-manganese dry batteries, but its capacity is more than twice that of the latter, so it has attracted people's close attention once it came out. Zinc-air batteries were mass-produced during World War I, but had a very low discharge current density of about 0.3 mA cm^{-2}. At that time, France applied them in railways, post and telecommunications and other fields, but has not fully demonstrated their superiority.

By the 1920s, a lot of research and improvement had been done on zinc-air batteries, and the direction had been shifted to alkaline zinc-air batteries. In 1932, Heise and Schumadcher changed the slightly acidic NH_4Cl electrolyte to alkaline to improve the conductivity of the electrolyte, which significantly reduced the internal resistance of the battery. It used zinc amalgam as the anode, porous carbon treated with paraffin waterproofing as the cathode, and 20% sodium hydroxide aqueous solution as the alkaline electrolyte, which greatly improved the discharge current, and the current density can reach $7 \sim 10$ mA cm^{-2}. This zinc-air battery exhibited high energy density, but low output power, and was mainly used for the power supply of railway signal lights and beacon lights. In the 1940s, due to the successful development of zinc-silver batteries, it was found that powdered zinc electrodes in alkaline solutions could discharge under high current conditions, which provided conditions for the further development of zinc-air batteries. Subsequently, the development of zinc-air batteries encountered a bottleneck for a long time due to the lack of suitable air electrode structures and effective catalysts.

After the 1960s, owing to the rapid development of normal temperature fuel cells, the research enthusiasm for high-performance gas electrodes was stimulated, which brought an opportunity to the development of zinc-air batteries and made another breakthrough in their performance. After 1965, the excellent gas diffusion electrode, prepared with PTFE as a binder and a hydrophobic agent, gradually replaced other gas electrodes. Such new type of gas diffusion electrode had a good gas/liquid/solid three-phase structure. Its electrode thickness was $0.12 \sim 0.5$ mm, and the discharge current density could reach 1000 mA cm^{-2} in pure oxygen state. In 1967, the above electrodes were further modified with a breathable and waterproof film to form a gas diffusion air electrode of a fixed reaction layer, so that they can work under normal pressure. Since then, commercialized zinc-air batteries have entered the market. By the end of the 1960s, high-efficiency zinc-air batteries had entered the stage of industrial production. At present, zinc-air batteries have played an irreplaceable role in many important fields.

1.2 Other Metal-Air Batteries

Metal-air batteries use relatively active metals such as magnesium, aluminum, zinc, cadmium, and iron as the anode, cooperating with the air electrode of the fuel cell, which uses the oxygen in the air as the cathodic active substance. The electrolyte generally uses alkaline or neutral electrolyte aqueous solution. If using more active lithium, sodium, potassium, etc. as anodes, it is necessary to match with non-aqueous organic or inorganic electrolyte solution. This kind of battery can be used as both a primary battery and a secondary battery. Due to its abundant raw materials, high cost performance, and pollution-free, metal-air batteries are called "green energy for the twenty-first century".

Since the air used in the air electrode is available everywhere and inexhaustible without cost, it is highly valuable to match with many metals with negative potential to form a series of metal-air batteries. The main feature of metal-air batteries is high specific energy [3]. In theory, the capacity of the cathode is infinite, and the active substance is outside the battery, which makes the theoretical specific energy of air batteries much larger than that of general metal oxide electrodes. The theoretical specific energy of metal-air batteries is generally above 1000 Wh kg^{-1}, and the actual specific energy is above 100 Wh kg^{-1}, belonging to high-energy chemical power sources. In addition, metal-air batteries are made of common and cheap materials and exhibit stable performance. If pure oxygen is used instead of air, the discharge performance can be greatly improved.

Metal-air batteries can be designed as primary batteries, reserve batteries, electrochemically rechargeable batteries, and mechanically rechargeable batteries. Metal-air batteries also have inherent shortcomings, that is, the battery cannot be sealed, which is easy to cause the electro-hydraulic to dry up and swell, and affects the capacity and life of the battery. If using alkaline electro-hydraulic, there exists carbonation effects, which increases the internal resistance of the battery and affects the discharge of the battery. In addition, wet storage performance is poor because the diffusion of air in the battery to the cathode will accelerate the self-discharge of the negative electrode. In order to reduce the self-corrosion of zinc, whether in alkaline or acid dry batteries, mercury or mercury compounds are generally added during production. However, mercury not only harms the health of workers but also pollutes the environment, so it needs to be replaced by a non-mercury corrosion inhibitor.

In addition to the zinc-air batteries introduced earlier, common metal-air batteries include magnesium-air batteries, aluminum-air batteries, and lithium-air batteries. Figure 1.1 shows a comparison chart of the theoretical mass energy density, volume energy density, and theoretical voltage of several metal-air batteries [4].

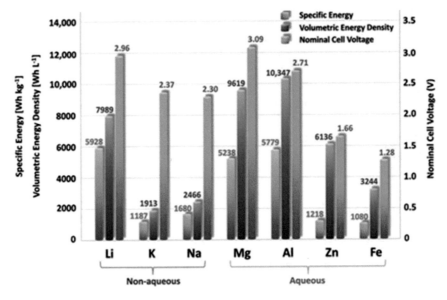

Fig. 1.1 Comparison of theoretical mass energy density, volume energy density, and theoretical voltage of several metal-air batteries [4]

1.2.1 Magnesium-Air Battery

Magnesium-air batteries use metal magnesium or magnesium alloys as the anodic active substance, and oxygen in the air as the cathodic active substance.

1. Overview of magnesium-air batteries

    Magnesium and magnesium alloys are excellent anode materials for metal-air batteries, with large crustal reserves, high stability to air and moisture, low density, high theoretical specific capacity, and high energy density [5]. Compared with the overly active lithium mental, the electrode potential of magnesium is relatively negative, which makes it more suitable for battery systems with electrolyte than lithium, also reduces the difficulty of production and storage, and no dendrites are generated during the cycling. Furthermore, the usage of magnesium not only reduces the cost, but also improves the safety [6]. Magnesium with low electrochemical equivalent can be paired with an air electrode to form a magnesium-air battery. The electrochemical equivalent of magnesium is $0.454 \text{ g (Ah)}^{-1}$, and $\varphi^{\ominus} = -2.69 \text{ V}$ in alkaline solution. The theoretical specific energy of the magnesium-air battery is 3910 Wh kg^{-1}, which is three times higher than that of the zinc-air battery.

    Although magnesium-air batteries have these advantages, there are still several unsolved problems that limit their wide applications [7–9]. First, magnesium is used as the anode of the air battery, and self-corrosion occurs in the electrolyte. The loss of this part of the electrode cannot be used for discharge, which reduces its anode efficiency; second, the magnesium anode will have undissolved particles fall off during the discharge process (called block effect). This weight loss also cannot be used for discharge, and it will also reduce its anode efficiency; third, many corrosion products will be formed on the surface of the magnesium anode during the discharge process. The corrosion products cover the anode surface and reeduuce the effective discharge area, which will hinder the further discharge and affect the discharge performance. Therefore, proposing new type of magnesium anode material with good corrosion resistance and stable discharge has become a key issue in the development of new energy sources. Alloying is a very effective solution. The second phase is one of the important factors affecting the anodic corrosion behavior and discharge performance of magnesium alloys. Therefore, the second phase can be adjusted by alloying to improve the performance of magnesium alloy anodes. Appropriate addition of zinc and calcium to magnesium can obtain good corrosion performance. In addition, there are also ways such as plastic deformation and heat treatment.

2.   Structure and working principle of magnesium-air battery

Magnesium-air batteries are generally rectangular in structure, with air electrodes on both sides of the battery, magnesium electrodes in the middle, and a separator between the positive and negative electrodes, and KOH solution as electrolyte.

The magnesium-air battery in alkaline solution reacts as follows during discharge:

$$\text{Negative electrode} : \text{Mg} - 2e + 2OH^- \rightarrow \text{Mg(OH)}_2 \qquad (1.1)$$

$$\text{Positive electrode} : \frac{1}{2}O_2 + 2e + H_2O \rightarrow 2OH^- \qquad (1.2)$$

$$\text{Cell reaction} : \text{Mg} + H_2O + O_2 \rightarrow \text{Mg(OH)}_2 \qquad (1.3)$$

Since magnesium-air batteries use sheet-like magnesium electrodes, the vacuum surface area is much smaller than that of porous powdered Zn electrodes in zinc-air. Therefore, when the same current passes through, the current density of magnesium electrodes is large, and the polarization is serious. At the same time, from the reaction formula, magnesium-air battery consumes water and it needs more electrolyte than Zn-air battery. For Zn-air battery, electro-liquid mass is 30 ~ 35% of the mass of active substance, while magnesium-air battery is 80 ~ 85%. The open circuit voltage of the magnesium-air battery is 1.6 V, and it can work between − 26 ~ 85 °C although the corrosion reaction is serious when discharging at high temperature. For example, only 40% of the rated capacity can be released at 52 °C. When discharging current density is higher than 40 mA cm^{-2}, the electro-hydraulic needs to be cooled. When working at low temperature, it can be changed to a neutral electrolyte, releasing 33% of its room temperature capacity at − 26 °C with electrohydraulic $NH_4Cl + CaCl_2$.

Due to the small electrochemical equivalent of magnesium and the high electromotive force of the battery, its theoretical specific energy is high, but in fact magnesium electrodes are easily passivated in an alkaline medium, so the actual specific energy is much lower than the theoretical value. It is of great significance for the development of magnesium secondary batteries to control the interlayer spacing of new layered materials and realize new layered cathode materials with high specific capacity, excellent rate performance and cycle stability [10].

1.2.2 Aluminum-Air Battery

The aluminum-air battery uses high-purity aluminum (Al) as the anode, oxygen as the cathode, and potassium hydroxide or sodium hydroxide aqueous solution as the active substance of the electrolyte.

1. Overview of aluminum-air battery

   Aluminum-air batteries have very high energy density and consist of an air cathode, an electrolyte, and a metallic aluminum anode (Fig. 1.2) [11]. The theoretical specific energy reaches 800 Wh kg^{-1}. Aluminum, rich in raw materials, is harmless to the human body, and can be recycled with no pollution to the environment, and the method of replacing aluminum electrodes can solve the problem of slow charging of aluminum-air batteries. Therefore, aluminum-air batteries have also received a lot of attention. Low-power aluminum-air batteries have been used in mining lamps, radio stations, and marine lighthouses. Developed countries led by the United States and the United Kingdom have carried out a lot of research and development on aluminum-air batteries. In the 1980s, The United States Aluminum Power Company studied the application of aluminum air battery in the power supply of deep-sea vehicle, unmanned exploration vehicle and AIP submarine. Using alloy Aluminum electrode and high efficiency air electrode, aluminum-air batteries have a specific energy of 400 Wh kg^{-1} and a power of more than 20 W kg^{-1}, and their energy density and volumetric energy density are several times higher than those of Cd-Ni batteries. The U.S. Department of Energy cooperated with Lawrence Livermore Laboratory in California to develop metal-air batteries for electric vehicles. Later, Lawrence Livermore Laboratory and other companies jointly established Voltek, firstly applying aluminum-air batteries to electric vehicles, and the efficiency of the aluminum-air battery pack is more than 90%. In 2015, Alcoa and Israel's Phinergy demonstrated that a 100 kg heavy aluminum air battery can drive a racing car for 1600 km.

   However, there are also some problems to be solved in the aluminum electrode of aluminum-battery. First, when the anode is dissolved, a passivation film is formed on the surface of metal aluminum, which inhibits the electron-loss oxidation reaction of aluminum, thereby increasing the potential of the aluminum electrode and causing the voltage drops of battery. Secondly, after the oxide film on the surface of aluminum is destroyed, a large amount of hydrogen will

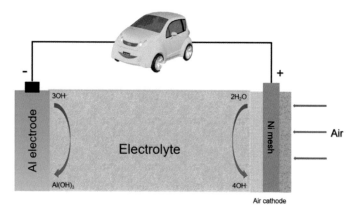

Fig. 1.2 Schematic diagram of aluminum-air batteries

evolve, and the dissolution of aluminum itself is difficult to stop, which eventually leads to serious battery self-corrosion discharge. Furthermore, commercial ultra-pure aluminum containing trace elements with low hydrogen evolution overpotential such as copper, iron and silicon will also aggravate the self-corrosion of aluminum. For example, the hydrogen evolution overpotential of copper is 0.59 V, and the hydrogen evolution overpotential of iron is 0.41 V [12]. Moreover, the metal-air battery represented by aluminum-air battery has a high theoretical energy density, but in the non-working state such as storage and standby, the anode often undergoes irreversible corrosion, resulting in the loss of energy density and the decrease of battery life. At present, there are many strategies to inhibit the corrosion of the anode of aluminum-air batteries, including anode alloying, electrolyte additives, gel electrolytes, non-aqueous electrolytes, etc. These strategies did alleviate the corrosion of the anode to a certain extent, but mostly at the expense of reducing the power density and energy density, and it is still necessary to find more effective strategies to improve the electrochemical performance of the system.

2. Structure and working principle of aluminum-air battery

The anodic active material of the aluminum-air battery is aluminum or aluminum alloy, and the cathodic active material of the is oxygen in the air. The aluminum electrode can use a neutral (salt-containing) solution or an alkaline solution, depending on the acidity and alkalinity of the electrolyte medium. The principles of electrochemical reactions are as follows:

1. Under neutral conditions

$$\text{Negative reaction:Al} \rightarrow \text{Al}^{3+} + 3\text{e}^- \tag{1.4}$$

$$\text{Positive reaction}: \text{O}_2 + 2\text{H}_2\text{O} + 4\text{e}^- \rightarrow 4\text{OH}^- \tag{1.5}$$

$$\text{Battery reaction}: 4Al + 3O_2 + 6H_2O \rightarrow 4Al(OH)_3 \qquad (1.6)$$

2. Under alkaline conditions

$$\text{Battery reaction}: 4Al + 3O_2 + 6H_2O \rightarrow 4Al(OH)_3 \qquad (1.7)$$

The actual electrode potentials of aluminum electrodes and air electrodes deviate greatly from the theoretical electrode potentials, and water is consumed during the reaction, but the actual energy density of aluminum-air batteries still exceeds that of most battery systems. In aqueous electrolyte, aluminum anode is prone to self-discharge and regassing. As a result, aluminum-air battery sources are often added with electrolyte just before use, or designed to mechanically rechargeable batteries that replace the aluminum anode after each discharge. Aluminum-air batteries emit a lot of heat when discharging, and the battery heats up so badly that a thermal management system is needed to prevent damage. At present, researchers are also exploring the possibility of combining oil and electrolyte to reduce the corrosion of the electrode and improve the performance and stability [13].

1.2.3 Lithium-Air Battery

Lithium-air battery uses lithium metal as the anode and oxygen in the air as the active substance of the cathode reactant.

1. Overview of lithium-air batteries
    Lithium-air batteries use metal lithium with minimum density and maximum electronegativity as the anode, porous air electrode as the cathode, and oxygen in the air as the anodic active material. Usually, a catalyst is used to promote the oxygen reduction reaction of the cathode. Therefore, the most prominent advantage of lithium-air batteries is that they have a very high theoretical energy density, with a theoretical specific energy of 11,140 Wh kg^{-1} (based on the mass of lithium metal), including the reaction product Li_2O_2, whose theoretical specific energy is 3505 Wh kg^{-1}. It has been considered as the most promising secondary battery system with an actual specific energy expected to reach 600 Wh kg^{-1}. With the development of the electric vehicle industry, more and more attention has been paid to power batteries. As a new type of chemical power system, lithium-air batteries are easier to meet the cruising range requirement of electric vehicles than other chemical power systems due to their outstanding energy density advantages. However, there are still many problems in current lithium-air batteries, such as high charge–discharge overpotential, poor cycle and rate performance, especially the metal lithium anode is faced with the fact that metal lithium is easily decomposed by various substances, resulting in many by-products, and using metallic lithium as the anode, lithium dendrites are generated during the

repeated electroplating/stripping of lithium ions. Therefore, metal lithium-air batteries still need to be further explored.

Lithium-air batteries mainly include battery types such as aqueous systems, organic systems, solid electrolyte systems and hybrid systems. Aqueous lithium-air batteries use a protective lithium metal composite anode and an aqueous electrolyte, which can work in an air environment, and the discharge products are usually LiOH or LiOAc. In the solid-state lithium-air battery, a solid-state electrolyte is used to separate the air electrode and the lithium anode, which avoids the direct reaction of moisture in the air with lithium metal, so that the battery can operate in the air. In theory, the solid-state electrolyte system battery can fundamentally solve the issues of security and stability. However, the lithium-ion conductivity of solid-state electrolytes is generally lower than that of liquid aqueous and non-aqueous electrolytes, the interfacial impedance between lithium metal and cathode is larger, resulting in the relatively lower energy utilization efficiency and output power of solid-state lithium-air batteries. At present, more research focuses on non-aqueous system (organic system) lithium-air battery, which was first reported by K. M. Abraham and Jiang in 1960.

2. Structure and working principle of lithium-air battery

Non-aqueous lithium-air batteries are mainly composed of metal lithium anode, organic electrolyte, and air electrode. During the charging and discharging process, the anode is mainly based on oxygen reduction (ORR) and oxygen evolution (OER) reactions, and the anode is based on the dissolution and deposition of lithium. During discharge, the lithium anode dissolves and transforms into Li^+, which migrates to the cathode (air electrode) through the electrolyte, and the electrons flow to the cathode through the external circuit to supply power to the load; meanwhile, oxygen is reduced at the cathode and combines with lithium ions to form Li_2O_2, which is deposited inside the porous air electrode.

The reaction mechanism of the lithium-air battery during the discharging process is as follows:

$$Negative : 4Al + 3O_2 + 6H_2O \rightarrow 4Al(OH)_3 \tag{1.8}$$

$$Positive : Li^+ + e + O_2 \rightarrow LiO_2 \tag{1.9}$$

$$LiO_2 + Li^+ + e^- \rightarrow Li_2O_2 \tag{1.10}$$

$$2LiO_2 \rightarrow Li_2O_2 + O_2 (chemical\ process) \tag{1.11}$$

$$Li_2O_2 + 2Li^+ + 2e^- \rightarrow 2Li_2O \tag{1.12}$$

The specific way that LiO_2 generates Li_2O_2 mainly depends on the properties of the electrode and electrolyte system. Theoretically, both Li_2O and Li_2O_2 may be the cathode discharge products of lithium-air batteries, but the main product in the actual system is still Li_2O_2. Under certain conditions, a small amount of Li_2O may be produced by a four-electron reaction pathway. The lithium-air battery system based on the four-electron reaction pathway can increase the theoretical energy density by more than 50% compared with the conventional system based on the two-electron reaction mechanism. In organic phase lithium-air batteries, the discharge product is solid Li_2O_2. Due to its own insulating properties, the cathode exhibits slow reaction kinetics, low energy efficiency, poor rate performance, and limited cycle stability. Similar effects can be weakened by controlling the nucleation and growth of Li_2O_2, such as the morphology and structure of Li_2O_2. Structural doping is an effective route to adjust the properties of materials, which can change the density of states, band gaps, and electrical conductivity of compounds, and endow new compounds with unique properties. This is an ideal way to improve the electrocatalytic activity of lithium-air batteries. Designing efficient catalysts to catalyze the cathodic reaction according to the four-electron reaction pathway is also the optional way to greatly advance the further development of lithium-air batteries.

1.3 Fundamental Principles

The electrochemical reactions involved in zinc-air battery are three-phase reaction on air electrode and oxidation–reduction reaction of metal zinc on zinc anode. Zinc-air batteries can be divided into primary batteries and secondary batteries. The primary zinc-air battery only involves the discharge process of the battery, that is, oxygen reduction reaction (ORR) occurs in the positive air electrode, and metal zinc oxidation reaction occurs in the anode, which loses its use value after the discharge. Nowadays, it has been widely used in low-current portable wearable devices such as artificial hearing aids. With the serious consumption of energy and great pollution to the environment by traditional vehicles, new energy vehicles have attracted wide attention due to the excellent advantages of energy saving and cleanliness. The demand for new energy vehicles in the automobile industry is constantly rising, which further promotes the rapid development of rechargeable secondary zinc-air batteries. Besides the same discharge process as the primary zinc-air battery, the secondary zinc-air battery also involves a charging process, that is, oxygen evolution reaction (OER) occurs at the positive air electrode and reduction of metallic zinc occurs at the anode. The reaction mechanism is much more complicated than that of the primary zinc-air battery, and the requirements for the air electrode are also much more complicated. In the secondary zinc-air battery, the air electrode needs to have dual-function activity of catalyzing OER and ORR reactions simultaneously, and these two reactions are mutually inverse. The process of four-electron transfer is involved, but the reaction process is complicated, and the steps are tedious. Since the kinetic process is slow and a large energy barrier needs to be overcome in the reaction process, the requirements for the catalyst are very strict. Therefore, in order to further study zinc-air battery, it is of great importance to deeply understand the mechanism of zinc-air battery.

1.3.1 Zinc Anode Mechanism

The anode of zinc-air battery is mainly metal zinc, and the involved reaction process is mainly the reduction and oxidation of metal zinc, also with the side reactions such as corrosion of zinc anode by electrolyte. During the preparation of zinc anode, the corrosion of electrolyte and the growth of its own dendrite are often considered. In the reaction process, theoretically, the catalytic electrode will not be consumed, so the zinc anode of zinc-air battery will determine the output capacity of the battery, which is an important factor restricting the development of zinc-air battery. The specific capacity of zinc-air battery can be improved by increasing the amount of zinc anode. The degradation of the performance of most batteries after long-term charge–discharge cycles is ascribed to the degradation of the metal anode, not the air electrode. However, the passivation of zinc electrode in alkaline solution will prevent the normal reaction, while the corrosion of zinc electrode will occur in acidic solution. Only when it is in neutral solution, the effect on zinc electrode is relatively small. The most common electrolyte is alkaline solution because the activity of catalyst in alkaline solution is high.

In an alkaline solution, the reaction process on the anode of zinc-air batteries can be divided into charging process and discharging process. When discharging, an oxidation reaction occurs on the anode, and metal Zn is oxidized to Zn^{2+}. The reaction equations are,

$$Zn + 4OH^- \rightarrow Zn(OH)_4^{2-} + 2e^- \tag{1.13}$$

$$Zn(OH)_4^{2-} \rightarrow ZnO + H_2O + 2OH^- \tag{1.14}$$

$$Zn + 2OH^- \rightarrow ZnO + H_2O + 2e^- \tag{1.15}$$

When charging, a reduction reaction takes place on the anode of zinc, and Zn^{2+} is reduced to metal Zn. The reaction equation are,

$$ZnO + H_2O + 2OH^- \rightarrow Zn(OH)_4^{2-} \tag{1.16}$$

$$Zn(OH)_4^{2-} + 2e^- \rightarrow Zn + 4OH^- \tag{1.17}$$

$$ZnO + H_2O + 2e^- \rightarrow Zn + 2OH^- \tag{1.18}$$

The two reactions in the charging and discharging process are inverse reactions to each other. However, in the primary battery, there is no reversible process, which is unique to the secondary battery.

1.3.2 Air Cathode Mechanism

The electrode is the core of zinc-air battery, which plays a vital role in the charging and discharging process of zinc-air battery. Oxygen precipitation reaction (OER) and oxygen reduction reaction (ORR) on the electrode belong to three-phase interface reactions, which occur at the interface of solid, liquid and gas, and are two indispensable reactions of zinc-air battery. In the secondary zinc-air battery, the ORR occurs on the air battery during discharge, while the OER occurs on the air electrode during charging, which requires that the air electrode has a higher reaction rate in ORR and OER. The catalysts on the air electrode are required to have good bifunctional activity and be able to reduce the energy barriers of the two reactions simultaneously. Therefore, the improvement of catalysts activity has now become a focus of research. Air electrode is often regarded as the energy converter of zinc-air battery, which determines the power output capacity of the battery. In the zinc-air battery, the process of three-phase reaction at the air electrode is as follows: when discharging, the oxygen in the air first diffuses through the diffusion layer of the air electrode to the catalytic layer, and then a three-phase electrochemical reduction reaction occurs at the interface between the catalytic layer and the electrolyte, just like the cathode process of hydrogen fuel cell.

OER is an anode reaction in water electrolysis for hydrogen production and air electrode charging in a reversible metal-air battery. During OER, oxygen molecules are produced by several proton/electron couplings. In order to produce oxygen molecules, a process with four electron transfers is required. However, the OER process is carried out through multi-step reactions in each step, so the potential energy barrier accumulated at each individual step will lead to the slow kinetics of OER reaction, resulting in a large overpotential. Different materials or crystal planes in the OER reaction will affect the formation of reaction intermediates, and then affect the reaction path. Figure 1.3 shows the mechanism of oxygen evolution reactions in acidic and alkaline electrolyte, respectively.

In alkaline electrolyte, the following two intermediate reactions first occur:

$$OH^- + * \rightarrow OH^* + e^- \tag{1.19}$$

$$OH^* + OH^- \rightarrow O^* + H_2O + e^- \tag{1.20}$$

Then, O_2 can form in two possible ways. One is to combine two O* intermediates to produce O_2:

Fig. 1.3 Mechanism diagram of oxygen evolution reaction in acidic (blue line) and alkaline (red line) electrolytes [14]

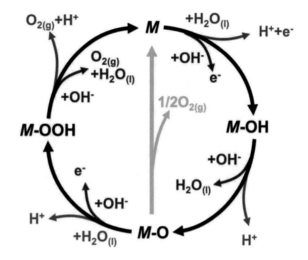

$$2O^* \rightarrow O_2 \qquad (1.21)$$

The other is that firstly O* reacts with OH^- to form OOH*, and then OOH* further combines with OH^- to form O_2 (therefore, the potential energy barrier of the reaction here is relatively large):

$$O^* + OH^- \rightarrow OOH^* + e^- \qquad (1.22)$$

$$OOH^* + OH^- \rightarrow O_2 + H_2O + e^- \qquad (1.23)$$

In acidic electrolyte, the reaction process is as follows:

$$H_2O + * \rightarrow OH^* + H^+ + e^- \qquad (1.24)$$

$$OH^* \rightarrow O^* + H^+ + e^- \qquad (1.25)$$

$$O^* + H_2O \rightarrow OOH^* + H^+ + e^- \qquad (1.26)$$

$$OOH* \rightarrow * + O_2 + H^+ + e^- \qquad (1.27)$$

In which "*" represents the active sites on the catalyst surface, and O*, OH* and OOH* represent the adsorbed intermediates. It can be seen from the above reaction process that the formation and dissociation of intermediate adsorption products O*, OH* and OOH* run through the whole reaction path, so the activity of OER catalyst can be described by the adsorption oxygen reduction free energy.

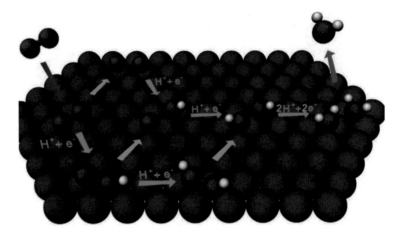

Fig. 1.4 Reaction path of ORR [15]

ORR is the cathode reaction of air electrode in metal-air battery, which will generate a variety of intermediate oxygen-containing species such as O^{2-}, OH^-, HO^{2-} and H_2O_2. Nowadays, the acceptable ORR reaction models are mainly divided into two ways: two electrons and four electrons, and the ORR process in different media systems will also have certain differences. The specific reaction mechanism is shown in Fig. 1.4.

In acidic electrolyte system, O_2 is first reduced to an intermediate product in the form of H_2O_2 by a two-electron pathway, and then H_2O is reduced from H_2O_2 by a two-electron pathway.

$$O_2 + 2H^+ + 2e^- \rightarrow H_2O_2 \qquad (1.28)$$

$$H_2O_2 + 2H^+ + 2e^- \rightarrow 2H_2O \qquad (1.29)$$

In addition, O_2 can be reduced to H_2O in one step by following the four-electron route without producing H_2O_2.

$$O_2 + 4H^+ + 4e^- \rightarrow 2H_2O \qquad (1.30)$$

In alkaline electrolyte system, O_2 is first reduced to HO^{2-} and OH^- by two-electron pathway, then HO^{2-} is reduced to OH^- by two-electron pathway.

$$O_2 + H_2O + 2e^- \rightarrow OH^- + HO^{2-} \qquad (1.31)$$

$$HO^{2-} + H_2O + 2e^- \rightarrow 3OH^- \qquad (1.32)$$

In addition, O_2 can be reduced to H_2O in one step by following the four-electron route without producing HO^{2-}.

$$O_2 + 2H_2O + 4e^- \rightarrow 4OH^- \tag{1.33}$$

From the above reaction process, it can be seen that the four-electron pathway is the most favorable for ORR regardless of the medium used. But the ideal four-electron pathway is usually difficult to occur. ORR usually follows a two-electron reaction path or a mixture of two-electron and four-electron reaction paths. Therefore, designing and preparing ORR electrocatalysts with high activity to promote the occurrence of four-electron reaction paths has become the focus of research.

1.4 Evaluation Elements

The performance of zinc-air battery can be evaluated from three aspects: OER, ORR and zinc-air battery, so as to determine whether zinc-air battery is suitable for large-scale application.

1.4.1 Evaluation Elements for OER

The OER performance is mainly evaluated from the aspects of initial potential, over-potential, Faraday efficiency, electrochemical active area, Tafel slope and stability, and they are defined as follows.

Initial potential: The initial potential refers to the electrode potential possessed when OER occurs. There are two ways to determine the value of the initial potential. First, after the nitrogen background is deducted from LSV curve, the potential deviating from the nitrogen background is the initial potential. Second, a relatively small current density is regarded as the initial potential, such as the potential corresponding to 0.1 mA cm^{-2} or 5% of the limit current density, which is regarded as the initial potential.

Overpotential: The difference between the applied potential and the equilibrium potential (1.23 V) of OER at the current density of 10 mA cm^{-2} is the overpotential, where 10 mA cm^{-2} represents the current density when the efficiency of a unit sunlight intensity in photolysis water reaches 10%.

Faraday efficiency: Faraday efficiency describes the electron transfer efficiency involved in OER. It can be calculated by comparing the actual gas production obtained in the experiment with the theoretically known gas production. The amount of gas O_2 produced can be determined by gas chromatography (GC) measurement or water displacement method. Faraday efficiency = (number of moles of actual product × number of reaction electrons × Faraday constant)/(current × time). In practice, not all current will be used for the catalytic reaction of OER, and other by-products may also occur, such as the oxidation of catalyst at high potential or the corrosion and decomposition of carbon-based substrate, which will generate new gas, so it is necessary to measure the actual generated gas.

Electrochemical active area: The actual electroactive area of the electrocatalyst can be obtained by calculating the electric double layer capacitance. In the experiment, the electric double layer capacitance can be measured by cyclic voltammetry.

Tafel slope: Tafel curve can be obtained from LSV curve of catalyst. It reflects the electrochemical kinetic performance of the catalyst. In OER polarization curve, the current density is closely related to the applied overpotential, which can be explained by Tafel equation: $\eta = a + b\log j$. Where η stands for over potential, b for Tafel slope, j for current density. When the value of η is zero, the corresponding value of j obtained from Tafel equation is the exchange current density (j_0), which is another criterion for evaluating the intrinsic activity of electrocatalysts. The ideal electrocatalyst should have smaller b and larger j_0.

Stability: The stability of OER catalyst is an important influencing parameter if it hopes to be used commercially on a large scale in industry. Stability can be used to evaluate the ability and state of catalyst for long-term working conditions. There are two ways to evaluate the stability of catalyst. One is the chronopotentiometry or chronoamperometry. If the catalyst has good stability, it shows extremely small fluctuation of potential or current density when running for a long time. Another judgment method is multi-cycle cyclic voltammetric test, that is, after thousands of CV, and continuing the LSV test, and observe the LSV polarization curves before and after scanning, then compare whether the curves before and after cycling are consistent to judge the stability of catalysts.

1.4.2 Evaluation Elements for ORR

There are many evaluation elements of ORR similar to OER, such as initial potential, electrochemical impedance, stability, Tafel slope, etc. ORR also has some unique evaluation elements.

Half-wave potential: The half-wave potential is generally the potential corresponding to the current which is half of the limit current density. Sometimes the potential corresponding to the current density of 3 mA cm^{-2} is taken as half-wave potential, because the limiting current density of ORR is 6 mA cm^{-2} without diffusion control and taking half of its value is 3 mA cm^{-2}. The higher the half-wave potential, the better the catalytic performance.

Transferred electrons number: ORR is a two-electron or four-electron transfer reaction, and the number of transferred electrons in ORR can be analyzed by fitting the polarization curves at different speeds. The ring-disk electrode is used to test the number of transferred electrons and the conversion rate of H_2O_2. Ring-disk electrode is a kind of rotating disk electrode where a layer of metal platinum ring is added on the periphery of glassy carbon electrode of rotating disk electrode, and the two are separated by an insulating ring of polytetrafluoroethylene. During the test process, a catalytic reaction takes place on the disk electrode. If an intermediate product is generated, it can be thrown to the Pt surface of the outer ring by rotating speed. If an oxidation or reduction potential is applied to the Pt ring electrode, the intermediate products will be decomposed, and the electrochemical workstation will be used to collect the generated current to judge the reaction path of the catalytic process.

1.4.3 Evaluation Elements for Zinc-Air Battery

The evaluation elements of Zinc-air battery are slightly different from those of OER and ORR, mainly from the aspects of open-circuit potential, charge–discharge polarization curve, maximum power density, constant current discharge, cyclic stability and so on.

Open-circuit potential: The potential measured when there is no discharge is the open-circuit potential. In fact, the open-circuit potential of zinc-air battery is about 1.40–1.50 V. The actual discharge potential is below 1.2 V.

Charge–discharge polarization curve: The charge–discharge potential is closely related to the charge–discharge current density. In the process of discharge, the discharge potential will gradually decrease with the increase of discharge current. The charging process is different from the discharging process, and the charging voltage will increase with the increase of the charging current density. The smaller the deviation of charging and discharging voltage from equilibrium potential, the smaller the polarization reaction of charging and discharging of zinc-air battery.

Maximum power density: Power is the energy released by the battery per unit time. In zinc-air battery, the product of potential and current density is plotted against current density, and the peak value of power density is the maximum power density.

Constant current discharge (specific capacity and specific energy): The specific capacity and specific energy of zinc-air battery can be calculated by discharging the battery at a constant current and recording the change of voltage, and then according to the quality of zinc consumed by the reaction.

Cyclic stability: Zinc-air battery has a stable charging and discharging platform, which can be tested for many times for a long time. The stability of catalyst and battery can be judged by the change of charging and discharging potential.

1.5 Composition of Zinc-Air Battery

Zinc-air battery is mainly composed of zinc electrode, separator, electrolyte and air electrode. Usually, the aqueous solution is potassium hydroxide, and all battery components are embedded in the casing. The zinc electrode is limited by a shell and a separator, and the separator is placed on top of the zinc electrode. The shell at the air electrode is open to the surroundings, usually equipped with air holes, so that

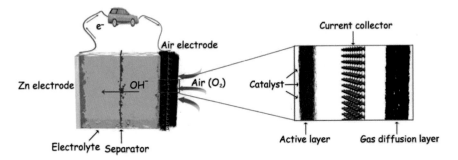

Fig. 1.5 Schematic diagram of zinc-air battery and its air electrode [16]

oxygen and the main reactants at the air electrode can enter. Figure 1.5 is a schematic diagram of a zinc-air battery and its air electrode.

1.5.1 Zinc Electrode

In zinc-air batteries, Zn is either applied in the form of particles or as a metal plate. Zn is filled into the shell in the form of particles together with liquid electrolyte and additives to form a paste. For the zinc-air battery equipped with this paste-like zinc electrode, usually 50% of the electrode volume is void and no active substance is filled [17]. Zinc electrode paste usually contains cellulose and surrounds by the zinc particles, to absorb the liquid electrolyte and ensure that the zinc electrode and the electrolyte are fully wetted.

The electrochemical stability of Zn is only below $- 0.98$ V versus SHE (standard hydrogen potential), so zinc will corrode when it encounters aqueous solutions such as KOH electrolyte [18]. This may lead to an unnecessary side reaction, that is, hydrogen evolution reaction (HER): $Zn + 2H_2O—Zn (OH)_2 + H_2$. This reaction has two negative effects on the operation of zinc-air battery. It consumes the active substance zinc and forms hydrogen, which will expand in the sealed zinc electrode, possibly leading to the structural change of the whole battery, or even the complete failure of the battery. Meanwhile, the molar volume of ZnO is 1.6 times that of Zn [19]. This shows that the bulk density of ZnO is 27% lower than that of Zn, and the particles in the zinc electrode will also experience significant volume expansion during discharge.

The chemical energy stored in the zinc electrode is directly converted into electric energy in the zinc-air battery, without intermediate steps of heat and mechanical energy conversion. The whole reaction process consists of oxidation reaction and reduction reaction. The energy storage device, determining the output capacity of the battery, needs to be able to maintain high activity and capacity during hundreds of cycles. The high specific capacity of zinc-air battery can be realized by increasing the amount of zinc anode. On the zinc anode, how to restrain dendrite growth, control oxygen evolution and improve the utilization rate of zinc electrode is the main problem to be solved.

1.5.2 Electrolyte

Ions generated and consumed in the battery need to be transported between the two electrodes to ensure the overall reaction. Electrolyte is the choice of ion transport medium. The electrolyte can be solid, gel polymer or liquid. In zinc-air battery, liquid electrolyte is usually used. The electrolyte is composed of a dissociative substance called solute on the one hand, and a solvent of solute substance on the other hand. In aqueous liquid electrolyte, water is the applied solvent. This solute is usually a dissociative salt, which can conduct ionic current for ion transmission between two electrodes.

Common liquid electrolytes are alkaline electrolytes, such as potassium hydroxide and sodium hydroxide. Among them, potassium hydroxide is more widely used than sodium hydroxide because of its higher solubility to zinc salt, better diffusivity to oxygen and lower viscosity. For example, the ionic conductivity of potassium ion (73.50 Ω^{-1} cm^2 equiv^{-1}) is much larger than that of sodium ion (50.11 Ω^{-1} cm^2 equiv^{-1}). At the same time, the experiment also proves that when platinum wire is directly used as air electrode for ORR test, potassium hydroxide is better than sodium hydroxide as electrolyte. The concentration of potassium hydroxide also plays an important role, but with the increase of the concentration, the viscosity of the electrolyte will greatly increase too, which seriously hinders the transmission of hydroxide ions. Experiments have proved that 6 M potassium hydroxide is suitable for zinc-air battery electrolyte. On the one hand, high concentration can provide high ionic conductivity and inhibit hydrogen evolution reaction of zinc anode. Furthermore, it has good transportation for oxygen and hydroxide ions. For secondary batteries, soluble zinc metal salts, such as zinc nitrate and zinc chloride, are generally added to meet the charging and discharging performance. Neutral electrolyte can reduce the corrosion of zinc anode and improve the stability of battery, but its catalytic activity is not as good as that in alkaline condition. However, the application of KOH electrolyte has some disadvantages, especially when compared with non-aqueous electrolyte: the water in aqueous electrolyte can evaporate, allowing the battery to eventually dry out. In addition, when contacting with carbon dioxide in the ambient air, alkaline KOH solution can also form carbonate species and lose hydroxide ions.

Gel electrolyte is often used to prepare flexible zinc-air batteries, which has different structures, such as sandwich structure, cable structure, page shape, etc. Besides the basic electrochemical properties, it also has the characteristics of folding, twisting, and stretching. The other is a cable-like flexible zinc-air battery. The metal is used as the central axis, and then it is wrapped by gel electrolyte. The air electrode wraps the outer side of electrolyte to form the basic structure of flexible battery, and finally the insulating material plays the role of packaging and protection at the outermost side.

1.5.3 Separator

The separator in the battery is a selective barrier to some of reactants and products of each electrode. For example, the gas O_2 cannot pass through the separator of zinc-air battery, otherwise it will be directly oxidized. More importantly, the separator needs to electrically isolate the two electrodes from each other and ensure sufficient OH^- ion transmission between the two electrodes. In addition, soluble intermediates from electrochemical reactions on zinc electrode and air electrode cannot pass through the separator, because they may cause side reactions or affect electrode reactions. However, Krej et al. found that the $Zn(OH)_4^{2-}$ ions could be transferred from zinc electrode to air electrode through separators commonly used in zinc-air batteries [20].

In general, separators are porous and very thin membranes. They consist of a polymer skeleton and certain additives. As batteries become more and more sophisticated, the available separators have a wide range of uses.

1.5.4 Air Electrode

The air electrode is the core of zinc-air battery, usually a gas diffusion electrode. The air electrode needs to satisfy the processes of air diffusion and oxygen oxidation and reduction. Oxygen is either consumed in the oxygen reduction reaction or produced in the oxygen evolution reaction. Generally, the air electrode consists of a current collecting layer, an active layer and a gas diffusion layer. The current collecting layer is sandwiched in the middle, which is the channel of electrons and the substrate with high conductivity, such as foam nickel, stainless steel mesh and carbon fiber cloth. The active layer is the side facing the electrolyte and consists of catalyst, carbon black and polytetrafluoroethylene (PTFE). The catalyst should reduce the polarization loss of the air electrode during the battery operation. As perovskite-based catalysts have the above two properties, they are usually used in rechargeable zinc-air batteries. The air diffusion layer is the side facing the air, which is the channel for oxygen transmission, ensuring a certain oxygen flow rate in the reaction zone of the active layer. It should have a high specific surface area with high porosity. The pores are partially filled with gas and partially wetted by electrolyte, which facilitates better oxygen diffusion. The high hydrophobicity is used to prevent the electrolyte from flooding the air electrolysis, which facilitates the adequate contact between the air electrode and oxygen.

1.6 Conclusions

As a clean energy that has received extensive attention, zinc-air batteries not only have obvious performance advantages such as high specific energy, low cost, and no pollution, but can be made into practical batteries of various types and specifications. In recent years, with the development of the manufacturing process, the continuous improvement of the theory and the battery's environmental adaptability, the product development of high-power zinc-air batteries gains a technical guarantee, which promotes the gradual commercialization of the zinc-air battery system. However, the practical application of zinc-air batteries still faces many challenges. The development of high-efficiency zinc-air batteries requires a reasonable understanding and design of electrodes, electrolytes, separators and other materials. Therefore, to overcome the key scientific and technological issues of Zn-air batteries and further promote their practical applications, it's necessary to gather collaborations between researchers from different backgrounds.

References

1. Z.W. Seh, J. Kibsgaard, C.F. Dickens, I. Chorkendorff, J.K. Nørskov, T.F. Jaramillo, Science **355**, eaad4998 (2017)
2. X. Li, H. Zhong, Q. Zheng, J. Yan, Y. Guo, Y. Hu, Bull. Chin. Acad. Sci. **34**, 443 (2019)
3. J. Fu, Z.P. Cano, M.G. Park, A. Yu, M. Fowler, Z. Chen, Adv. Mater. **29**, 1604685 (2017)
4. X. Li, H. Lu, S. Yuan, J. Bai, J. Wang, Y. Cao, Q. Hong, J. Electrochem. Soc. **164**, A3131 (2017)
5. L. Wang, D. Snihirova, M. Deng, B. Vaghefinazari, S.V. Lamaka, D. Höche, M.L. Zheludkevich, J. Power Sources **460**, 228106 (2020)
6. Y. Feng, W. Xiong, J. Zhang, R. Wang, N. Wang, J. Mater. Chem. A **4**, 8658 (2016)
7. X. Chen, Q. Zou, Q. Le, J. Hou, R. Guo, H. Wang, C. Hu, L. Bao, T. Wang, D. Zhao, F. Yu, A. Atrens, J. Power Sources **451**, 227807 (2020)
8. B. Xiao, G.L. Song, D. Zheng, F. Cao, Mater. Des. **194**, 108931 (2020)
9. N. Wang, R. Wang, C. Peng, B. Peng, Y. Feng, C. Hu, Electrochim. Acta **149**, 193 (2014)
10. L. Zhou, Q. Liu, Z. Zhang, K. Zhang, F. Xiong, S. Tan, Q. An, Y.-M. Kang, Z. Zhou, L. Mai, Adv. Mater. **30**, 1801984 (2018)
11. L. Fan, H. Lu, J. Leng, Z. Sun, C. Chen, J. Power Sources **299**, 66 (2015)
12. Y. Liu, Q. Sun, W. Li, K.R. Adair, J. Li, X. Sun, Green Energy Environ. **2**, 246 (2017)
13. B.J. Hopkins, Y. Shao-Horn, D.P. Hart, Science **362**, 658 (2018)
14. N.T. Suen, S.F. Hung, Q. Quan, N. Zhang, Y.J. Xu, H.M. Chen, Chem. Soc. Rev. **46**, 337 (2017)
15. W. Xia, A. Mahmood, Z. Liang, R. Zou, S. Guo, Angew. Chem. Int. Ed. **55**, 2650 (2016)
16. J. Pan, Y.Y. Xu, H. Yang, Z. Dong, H. Liu, B.Y. Xia, Adv. Sci. **5**, 1700691 (2018)
17. A.A. Mohamad, J. Power Sources **159**, 752 (2006)
18. J.S. Lee, S.T. Kim, R. Cao, N.S. Choi, M. Liu, K.T. Lee, J. Cho, Adv. Energy Mater. **1**, 34 (2011)
19. W.G. Sunu, D.N. Bennion, J. Electrochem. Soc. **1980**, 127 (2017)
20. I. Krejčí, P. Vanýsek, A. Trojánek, J. Electrochem. Soc. **140**, 2279 (1993)

Chapter 2
Cathode Materials for Primary Zinc-Air Battery

2.1 Overview

The operation of primary zinc-air batteries mainly depends on the ORR process of the air cathodes, so the key component of the air electrodes is the ORR electrocatalysts [1]. However, the slow kinetics of ORR leads to high overpotential, which reduces energy efficiency and ultimately limits the output performance of primary cells [2]. The performance requirements for efficient electrocatalysts include high active site density and uniform distribution for high ORR onset potential and high catalytic activity [3, 4]. The high surface area and sufficient porous structure provide adequate mass transfer pathways and enhanced electrode kinetics. Robust construction can produce chemical and mechanical stability, high durability, high quality, and volumetric activity. Meanwhile, it has abundant resources and low cost [5–7]. However, most electrocatalysts do not reach an acceptable level. Therefore, the actual energy density of existing zinc-air batteries is usually only 40–50% of the theoretical density.

In principle, two standard ORR pathways occur in alkaline solutions including the direct four-electron pathway (H_2O) and the indirect two-electron pathway (H_2O_2) [8]. In the former case, one oxygen molecule accepts four electrons and is reduced to four OH^- ($O_2 + 2H_2O + 4e^- \rightarrow 4OH^-$, 0.4 V). In the latter case, oxygen molecules are first reduced to intermediate H_2O_2 and further reduced to form OH^- ($O_2 + H_2O + 2e^- \rightarrow HO_2^- + OH^-$ (0.07 V), $HO_2^- + H_2O + 2e^- \rightarrow 3OH^-$ (0.87 V), (Fig. 2.1) [9]. The four-electron mechanism is more efficient and favorable for Zn-air batteries, which avoids premature degradation of electrochemical cells caused by the corrosion/oxidation of carbon scaffolds and other materials by peroxides. Benefiting from a similar principle, most oxygen catalysts used in alkaline fuel cells and other alkaline metal-air batteries can also be used in zinc-air batteries. A large number of materials have been intensively investigated in the field of ORR electrocatalysts, including noble metals and their alloys, transition metals, metal oxides/calcites/carbides/nitrides, composites, and carbon nanomaterials [10, 11].

© The Author(s), under exclusive license to Springer Nature Singapore Pte Ltd. 2023
S. Peng, *Zinc-Air Batteries*, https://doi.org/10.1007/978-981-19-8214-9_2

The primary zinc-air battery does not have a charging process because it only involves the discharge process during the service process. Therefore, the requirements for catalysts and electrolytes are lower than that of secondary zinc-air batteries and their commercial application is also less difficult. Meanwhile, the single discharge capacity of primary zinc-air batteries is not only much higher than that of lithium-ion batteries but also higher than that of secondary zinc-air batteries. Compared with the widely used lithium-ion battery, it has certain advantages. In practice, there have been commercialized primary zinc-air batteries with an energy density of 450 Wh kg^{-1}, which are mainly used in hearing aids, navigation buoys, railway signal lights, telecommunication devices, and electrocardiographic telemetry devices. However, the energy density of the commercialized zinc-air battery is still much lower than the theoretical value. Therefore, its application is limited to some devices with low power and continuous operation for a long time. To expand the application range of zinc-air batteries, it is necessary to improve the energy density and rate performance of zinc-air batteries, which puts forward higher requirements for cathode materials.

The huge specific energy, low cost, and safety of the primary zinc-air battery make it a worthy choice for some portable electronic products. It is particularly economical for applications where the battery energy is depleted within one to fourteen days. The enormous specific energy of the zinc-air battery can be released because the energy is depleted within one to fourteen days. The battery voltage is relatively gentle during most of the discharge process.

As described in Chap. 1, the oxygen reduction reaction (ORR) occurs at the positive electrode of a primary zinc-air battery. Since the ORR reaction involves multiple electron transfer steps and the adsorption and desorption process of a large number of reaction intermediates (-OOH, -O, -OH), the reaction process is kinetically slow and needs to overcome a large energy barrier in thermodynamics. It is only possible for ORR to meet the actual demand under the action of a catalyst. As a

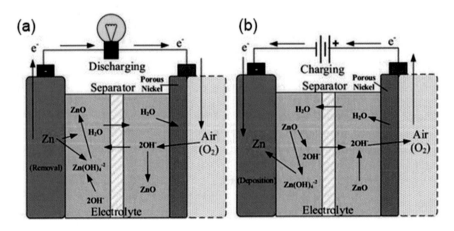

Fig. 2.1 Schematic diagram of a zinc-air battery in **a** discharged and **b** charged states [11]

result, the positive ORR catalyst has a great influence on the performance of the Zn-air battery, which largely affects the key parameters such as the operating voltage and discharge capacity of the Zn-air battery. Based on the above facts, this chapter mainly discusses only oxygen reduction catalysts and their applications in primary zinc-air batteries.

2.2 Noble Metal-Based Catalysts

Currently, precious metals including platinum, palladium, and gold are undoubtedly the most efficient ORR catalysts. Due to its scarcity and high price, the improvement of intrinsic activity and utilization is an important issue. It has been reported that modulation of electronic structure (d-band), morphology (surface strain, surface composition, and surface cross-section), and optimization of reactant adsorption sites (defect and size structure) are considered to be the main strategies to solve the above problems. Mass activity (MA) as assessed by specific activity and electrochemically active surface area (ECSA) are important performance parameters and indicators for optimizing ORR activity and reducing catalyst cost.

2.2.1 One-Component Precious Metal Catalyst

For single-component noble metals of Pt/Pd, MA can be improved by modulating the chemical environment and fine structure such as exposed catalytically active surface, morphology, particle size, and near-surface strain. Studies have shown that the ORR activity on different Pt (hkl) crystal surfaces increased in the order (100) < (111) < (110) [12]. Therefore, it is a satisfying strategy to develop noble metal catalysts with different morphologies (concave/convex polyhedra, nanowires, nanocages, and other complex structures) to expose specific facets, which can generate more unsaturated coordination and enhance ORR activity [13–15]. For example, the ORR activity of platinum concave nanocubes with (510), (720), and (830) high-index facets is significantly enhanced compared to other morphologies of platinum catalysts (cube, cuboctahedron) and commercial Pt/C catalysts with (100) and (111) low-index faces. Reducing particle size is another viable strategy to increase the number of accessible surface atoms to improve ORR activity. Furthermore, strain tuning can optimize the lattice edges of surface atoms and the reactivity-strain relationship can provide guidelines for tuning ORR activity [16, 17]. For example, shape-controlled ultrafine zigzag platinum nanowires with high surface strain and non-coordinated rhombohedral-rich surface configuration exhibited a competitive MA of 13.6 A mg_{Pt}^{-1}. The compressive strain favors weakened adsorption on the platinum surface to optimize ORR activity, but the tensile strain has the opposite effect (Fig. 2.2a, b) [18]. The deposited platinum nanofilms with compressive and tensile states successfully illustrate the different strain effects on platinum surface adsorption [19]. These different strain states cause

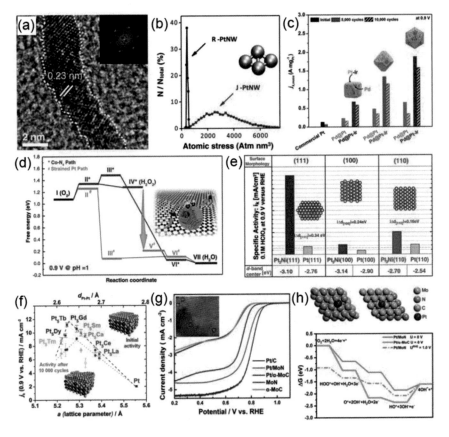

Fig. 2.2 **a** High-resolution transmission electron microscope (HRTEM) and corresponding FFT (inset) images, and **b** mean atomic stress of zigzag platinum nanowires [18]. **c** ORR activity of different catalysts at 0.9 V [20]. **d** Free energy diagram of the ORR pathway of platinum core–shell nanoparticles [21]. **e** Structure and ORR activity of Pt₃Ni(hkl) and Pt(hkl) surfaces [12]. **f** Volcano-type relationship between activity (recorded at 0.9 V), lattice parameters, and Pt-Pt distance of polycrystalline Pt₅M catalysts [22]. **g** ORR activity and **h** theoretical explanation of Pt/Mo$_x$ (X represents C or N) [23]

the shift of the D-band center of the platinum nanofilm, leading to changes in ORR activity.

2.2.2 Precious Metal Alloy Catalysts

Due to the similar properties of noble metals, the construction of Pd@M (M, Pd, Pt, and Au) with different surface structures is an effective strategy to optimize the ORR performance such as nondendritic and core–shell (cubic, octahedral, icosahedral, hollow, and nanocage) [24, 25]. DFT calculations suggest that the weakened EO on the M surface is responsible for the enhanced ORR activity of Pd@Mn [26]. For example, the constructed Pd@Pt core–shell decahedron exhibited enhanced ORR activity and durability. In addition, different exposed facets and surface compositions of Pd@Pt catalysts have also been extensively investigated. For example, hollow Pt-Pd nanocages with well-defined cross-sections (111 or 100) showed significantly improved ORR activity over solid Pd@Pt nanocrystals [27]. In subsequent work, the effect of Pd@PtnL core–shell ORR activity was investigated by depositing different numbers of Pt atomic layers on the Pd core [28, 29]. Specifically, Pd@PtnL octahedra tended to undergo mixed surface components due to the edge substitution of Pd by Pt, which could enable the presence of more favorable ORR energies on pure Pt shells than Pd-Pt shells. However, electrochemical selective leaching of the core Pd element in the accelerated durability test for the Pd@Pt core–shell nanocatalysts can lead to undesirable structural degradation and poor stability. Various Pt-Pd-Au alloy catalysts have been developed based on the tunable ligaments and pore sizes of nanoporous metals for efficient and stable ORR [30]. For example, an unsupported nanoporous catalyst with a Pt-Pd shell on Au was constructed, in which the initial MA could be increased in the first 30,000 cycles and even remained stable after 70,000 cycles [31]. Due to the thermodynamic stability and resistance to leaching of Ir during the ORR process, another research direction created Pd@Pt-Ir to enhance the ORR activity and durability [32, 33]. The catalytic properties of Pd@Pt-Ir nanocrystals are related to the enhanced ORR activity, which can be attributed to the easier protonation of O* and OH* under the cover of OH* after the incorporation of Ir atoms into the Pt lattice (Fig. 2.2c).

Alloying of noble metals (Pt/Pd) with transition metals (M=Ni, Co, Fe, Ti, V, Cu, etc.) is another effective strategy to improve its intrinsic activity towards ORR because it can reduce the ε_d of Pt to regulate oxygen adsorption energy of intermediate products on metal surfaces [34, 35]. It is generally believed that the d-band center can determine the surface adsorption properties of Pt-based alloys. Furthermore, ligand/strain effects can be introduced by combining other transition metal atoms with Pt [22]. DFT calculations show that PtM alloys have a volcano-type relationship as a function of the d-band center. Among them, the Pt_3Co, Pt_3Ni, and Pt_3Fe alloys located at the top of the volcano plot should exhibit efficient ORR activity [36]. Likewise, the MA of PtM alloys can be improved by tuning the chemical environment and fine structure, such as the exposed catalytically active surface, morphology, strain tuning, and electronic state, which can weaken the chemisorption of surface oxygen species and shift the d-band down [37–42]. For example, the synergy between the ultra-low-loaded Pt of strained Pt-Co core–shell nanoparticles and the metal-catalyzed substrate of the Pt-free group together contributed to the improved MA and ORR persistence of 1.77 A mg_{Pt}^{-1} (Fig. 2.2d). Compared with the Pt(111) surface, the enhanced ORR performance of the Pt_3Ni(111) surface was caused by the synergistic effect between the down-shifted d-band center (0.34 eV), ligands, and morphology, resulting in improved Pt utilization rate and adsorption strength (Fig. 2.2e) [12, 34]. As another example, a highly crystalline $PtNi_3$ nanoframe with an open structure showed sufficient exposure of Pt(111) and surface strain of Pt atoms, and the synergistic effects are beneficial to enhance ORR activity [43]. Furthermore, highly ordered alloys or intermetallic compounds with strong Pt-M interactions and increased Pt-M bonds can tune the local Pt atomic structure, resulting in more advanced intrinsic properties compared to disordered alloys [44–47]. Compared with Pt-rich ordered intermetallics and low Pt disordered alloy catalysts, this core–shell structured $PtCo_3$ ordered catalyst exhibited durable Pt surface compressibility and high ORR activity. Nonetheless, the leaching of active transition metals under realistic conditions (acidic corrosion environment at high potential/voltage) limits their persistence [48]. Alloying Pt with early transition metals (Y, Gd, Tb, etc.) with less electronegativity could effectively improve the ORR activity and stability by suppressing lattice atomic diffusion because they can donate electrons to reduce the ε_d of Pt [49, 50]. For example, various platinum-lanthanum alloy/early transition metal catalysts showed a volcano relationship between ORR activity and the lattice parameters of these metals (Fig. 2.2f) [22]. As another example, ultrathin PtGa alloy nanowires with unconventional p-d hybridization exhibited higher ORR performance and slight MA loss after 30,000 cycles of durability testing.

2.2.3 Noble Metal Single-Atom Catalysts

Single-atom catalysts (SACs) have unique electronic structures. The unsaturated coordination environment of the active centers, as well as the fully exposed active sites, can achieve 100% maximum efficiency showing good potential compared to their particle state. In particular, Pt-based SACs showed superior ECSA and competitive ORR activities at lower loadings. However, it is a challenge to fabricate SACs and keep metal species atomically dispersed with low-cost equipment and high yields.

Recently, various strategies such as wet impregnation, co-precipitation, atomic layer deposition, electrochemical displacement, photochemical reduction, and steric confinement have been developed to synthesize platinum-based SACs for ease of operation and possible large-scale fabrication [51, 52]. For example, Pt SAC was fabricated on N-doped carbon black support by the wet impregnation method, which showed high Pt utilization in fuel cells (0.13 g_{Pt} kW^{-1}) [53]. DFT calculations indicated that pyridine-N doped on carbon strongly anchored Pt single atoms for highly active ORR. To prevent the migration and aggregation of Pt SACs, the photochemical solid-phase reduction was used to synthesize atomically distributed Pt supported on N-doped porous carbon at 3.8 wt% [54]. The pyridine-like Pt-N_4 coordination could stabilize the Pt SAC and serve as the active sites, so the optimized chemical and electronic structure of Pt-N_4 could help to improve the ORR activity. Furthermore, the coordination environment of the metal center is an important component of the active site and plays a decisive role in determining ORR activity. For example, compared with Pt/α-MOC SAC (Pt atom coordinated to Mo), the synthesized Pt/MoN (Pt atom coordinated to N) possessed a 15-fold higher MA, which was similar to Pt loading (Fig. 2.2g) [23]. DFT calculations indicated that the coordination of Pt with N atoms modulated the electronic properties of Pt and induced more positive charges on Pt, resulting in relatively weaker OH* adsorption and better ORR performance (Fig. 2.2h).

In principle, the reduction of platinum nanoparticles to platinum single atoms is one of the most effective ways to improve platinum utilization efficiency and reduce the cost of ORR electrocatalysts [53]. However, it has also been reported that Pt single atoms may not be ideal ORR catalysts because the cleavage of the O-O bond usually occurs at co-located Pt sites hindering the 4-electron pathway on Pt SAECs [55, 56]. Liu et al. engineered defects on the carbon support to improve the ORR activity of Pt SAECs [57]. Structural characterization and DFT calculations indicated that single platinum atoms were anchored to four carbon atoms in divacancy graphene (denoted as g-2-Pt in their report), which were considered active sites. The presence of double vacancies made the platinum atoms almost in the same plane as the graphene framework, which promoted the adsorption of O_2 in the end-face form (Fig. 2.3c). A "downhill" path can be accommodated on g-2-Pt according to the calculated free energy diagrams of ORR on different substrates, resulting in enhanced ORR activity (Fig. 2.3d). This suggests that the rational design of Pt-based SACEs can still yield efficient and durable ORR electrocatalysts.

Among various catalysts, the adsorption free energy of atomic Ir-N-C configuration OH (ΔG_{OH}) is the closest to the peak of the classical volcano diagram, indicating high catalytic activity for ORR [58]. This theoretical prediction is consistent with iridium porphyrins exhibiting the highest ORR activity in acidic electrolytes [59]. Motivated by this prediction, Chen's group used ZIF-8 as a host for Ir precursors to obtain Ir-N-C SAECs for ORR [60]. STEM and XAS analysis demonstrated that the active site was a single Ir atom coordinated with four N atoms. As expected, the obtained Ir-SAEC exhibited remarkable catalytic performance for ORR in acid with $E_{1/2}$ as high as 0.864 V_{RHE}. DFT calculations suggested that a modest ΔG_{OH*} was responsible for the enhanced catalytic activity. This finding provides important guidance for the future development of high-performance Ir-based SAECs.

The ORR of Pd single-atom supported on graphite nitride (Pd/g-C_3N_4) was computationally investigated at different Pd coverages [61]. It was found that the ORR activity depended on the coverage of Pd, and SAEC with 25% coverage showed the lowest energy barrier for ORR at 0.39 eV, even lower than that of Pt (0.80 eV). Verification of this prediction has not been demonstrated experimentally, which can be a challenge due to the relatively low conductivity of the C_3N_4 substrate.

2.3   Transition Metal-Based Catalysts

Oxide/carbide/phosphide/sulfide/nitride/phosphide   and   transition   metal-based compounds are abundant and inexpensive in terms of ORR, making them viable alternatives to precious metals for large-scale applications in renewable energy technologies. However, compared with noble metal-based catalysts, they have poor intrinsic activity and cannot meet the needs of long-term operation. Recently, various strategies have been developed, including defect engineering, modulation

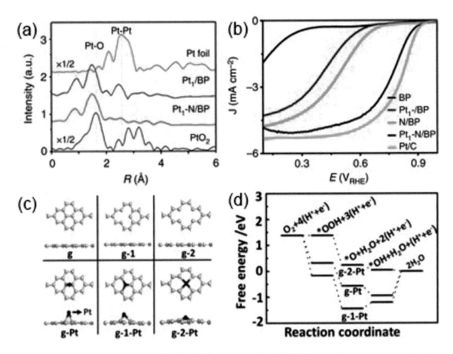

Fig. 2.3 **a** Platinum L3 edge XANES of different samples. **b** RRDE polarization curves of carbon black (BP), N-doped carbon black (N/BP), Pt1-BP, Pt1-N/BP, and commercial Pt/C in O_2 saturated electrolyte with a scan rate of 5 mV s^{-1} at a rotational speed of 1600 rpm [53]. **c** Optimized structures for different substrates: pristine graphene (g), single-voided graphene (g-1), two-voided graphene (g-2), single platinum atom supported on pristine graphene (g-Pt), single platinum atom Supported in single-voided graphene (g-1-Pt, Pt-C3), single platinum atoms supported in two-voided graphene (g-2-Pt, Pt-C4). Grey and black spheres represent carbon and platinum atoms, respectively. **d** Free energy diagrams for the complete reduction of O_2 on g-Pt, g-1-Pt, and g-2-Pt substrates at 0.83 V, respectively [57]

of electronic structure, and integration of conductive supports to improve their ORR activity.

2.3.1 Simple Transition Metal Oxide/Carbide Catalysts

Transition metal oxides have attracted great attention as efficient electrocatalysts for ORR. Transition metals are unique because they can form different cationic oxidation states. The diversity of their oxides allows them to be tuned with a high tolerance for better ORR performance. They are more readily available and less expensive than typical noble metal catalysts, making them more accessible for practical applications. However, there are still some restrictions on their development (Fig. 2.4). Here, we

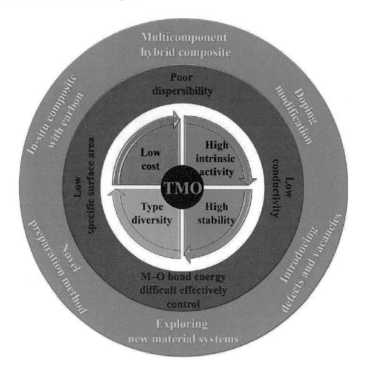

Fig. 2.4 Challenges and opportunities for transition metal oxides and common synthesis strategies [62]

define simple transition metal oxides as single metal oxides containing only one transition metal and oxygen.

In general, the ORR activity of transition metal (manganese, iron, cobalt, nickel, etc.) oxides is related to metal vacancies and metal–oxygen covalency [63]. For example, oxygen vacancies in MnO_2 can enhance the interaction of oxygen-containing species and lower the kinetic energy barrier, thereby enhancing the ORR performance. However, the low conductivity limits its ORR catalytic activity. The introduction of V with metallic features can tune the electronic structure and conductivity of Co oxides [64]. For example, spinel oxides (Co_2VO_4) have been synthesized by combining V atomic chains with electroactive Co cations [65]. Compared with pure Co oxide, the conductivity of Co_2VO_4 was improved by several orders of magnitude. Experimental and simulation analyses reveal that the Co(II) cations on the octahedral sites exhibited a low spin state with one eg electron, which was favorable for optimizing the thermodynamics of ORR. The combination with carbon is a common way to improve conductivity. For example, N-doped graphene supports exhibited competitive ORR activity with Co_3O_4 in alkaline solution due to the synergistic chemical coupling [66]. Likewise, a tubular monolayer superlattice of hollow Mn_3O_4 nanocrystals with a thin carbon coating showed competitive ORR performance for improved electron transport and enhanced ECSA [67].

Recently, metal centers and carbon are proposed as active sites in these carbides. Therefore, defect engineering of carbon, morphology, and strain tuning of carbides are effective strategies to improve their ORR performance. For example, tungsten carbide encapsulated in a high-defect graphite layer has a high density of sub-nanopores channels, which significantly improves the electrical conductivity, optimizes the penetration and exchange of different components, and effectively suppresses the agglomeration and coarsening of the catalyst during the ORR process [68]. As a new class of carbides, MXenes possess a high ECSA, excellent electrical conductivity, thermal stability, and strong acid resistance, which have great potential for electrocatalysis [69, 70]. In this regard, a series of MC_2 (TiC_2, VC_2, NbC_2, TaC_2, MoC_2) were investigated as electrocatalysts for ORR [71]. TaC_2 and MoC_2 exhibited higher ORR activity compared with Pt, which was ascribed to the existence of dual active sites. More importantly, this optimized active site can break the traditional single active site ratio relationship. Meanwhile, transition metal carbides containing two or more metals have higher ORR activity than single-metal carbides because of the synergistic effect and surface strain between each component [72]. As an example, the carbonized CoFe nanostructures encapsulated within N-doped graphene nanosheets synergistically provided more active sites and improved interfacial charge transfer accompanied by optimized oxygen adsorption energy [73].

2.3.1.1 Simple Manganese-Based Oxide Catalysts

Manganese oxides (MnO_x) for ORR have been studied since 1973 [74]. So far, more than 20 polycrystals with multivalent and non-compact properties have been studied due to the valence variation of manganese. The catalytic ability of manganese oxides for ORR depends on a variety of factors, including chemical composition, crystal structure, oxidation state, morphology, and surface area [75, 76].

With different valences of manganese, manganese oxides including Mn_5O_8, Mn_3O_4, Mn_2O_3, MnO_2, MnO, $MnOOH$, and amorphous MnO_x all exhibit electrocatalytic activity towards ORR in alkaline media. The electrocatalytic activity was reported to increase in the order of $MnO < Mn_5O_8 < Mn_3O_4 < Mn_2O_3 < MnOOH$. The activity of MnO_2 depends to a large extent on its crystalline phase [76–78]. $MnOOH$ is considered to have the highest ORR activity. This may be because many research works have found that the improved catalytic performance of MnO_x was related to the formation of $MnOOH$. Mn_3O_4 and Mn_2O_3 have also been extensively studied due to their considerable ORR activity. The Mn_3O_4 catalysts established only on glassy carbon supports could even show an improved ORR onset potential of 0.80 V and a four-electron process [79]. The nanowire-like Mn_2O_3 exhibited remarkable catalytic activity for ORR comparable to Pt/C in alkaline medium [80]. The unique nanostructure of Mn_2O_3 with a large aspect ratio and preferentially exposed (222) crystalline surface contributed to the enhanced ORR activity.

The crystal structure can also have a significant impact on ORR performance [76]. For example, the catalytic activity of manganese dioxide with different crystal structures (such as α, β, γ, λ, and δ) is in the order of β-manganese dioxide < λ-manganese dioxide < γ-manganese dioxide < α-manganese dioxide. The order of delta-manganese dioxide increases in the field of metal-air batteries [81]. Alpha and delta manganese dioxides are often highlighted for their excellent electrochemical activity in practical applications. Regarding the electrocatalytic activity, there is another view, that is, the order of δ-manganese dioxide < β-manganese dioxide < AMO < α-manganese dioxide [75]. The electrocatalytic performance of δ-manganese dioxide is not stable, possibly due to the combined effect of its inherent tunnel size and conductivity through different preparation processes. Due to the variability of manganese oxides, the order of their catalytic abilities remains controversial. For example, the electrochemical performance of λ-MnO_2 is inferior to that of α- and δ-MnO_2, while the highly crystalline λ-MnO_2 prepared by electrochemically removing lithium from $LiMn_2O_4$ is an efficient ORR electrocatalyst. By introducing oxygen vacancies, λ-MnO_2 exhibits a high diffusion-limited ORR current and a large electron transfer number (over 3.8) [82].

Manganese oxides exist in different morphologies, such as sea urchin-like [83], star-like [84], nanospheres [85, 86], nanowires [87, 88], nanorods [80], nanotubes [67, 82], nanosheets [89], nanoflowers [90], nanosheets [91], and mini-cubes [92], etc. Due to the complexity of manganese oxide, the relationship between the structure and ORR activity has not been consistently concluded. It has been reported that the electrocatalytic activity of MnO_2 depends on the nano shape and follows the order of nanowires > nanorods > nanotubes > nanoparticles > nanoflowers [90]. However, it has also been found that single-crystal-MnO_2 nanotubes exhibited better stability than α-MnO_2 nanowires and d-MnO_2 nanosheets as oxygen reduction catalysts in microwave-assisted hydrothermally synthesized air cathodes [89]. In addition, freestanding and ultrathin α-MnO_2 nanosheets also exhibited excellent ORR mass activity of 21 ± 1.2 mA mg^{-1} at 0.75 V against RHE [93]. There is also a structure–activity relationship for Mn_3O_4. The Mn_3O_4 nanosheets with preferentially exposed (001) crystal planes showed distinct ORR activity than the Mn_3O_4 nanorods with preferentially exposed (101) crystallographic planes. A unique, tailored, self-assembled, free-standing, carbon-coated tubular superlattice composed of the monolayer hollow Mn_3O_4 nanocrystals described in Fig. 2.5 was fabricated by a template-assisted epitaxial assembly strategy [67]. The unique layered structure not only improved the inherent low electrical conductivity and electrochemical structural stability of Mn_3O_4 but also increased the available active sites. The expected excellent ORR performance of the above catalysts is comparable to that of Pt/C and higher than that of most reported manganese oxides.

The excellent ORR performance is ascribed to the combination of favorable structural features, including tubular geometry, monolayer superlattice structure, and carbon coating. It is concluded that the ORR activity of manganese oxide can be enhanced by rationally designed structures. The Mn_2O_3 catalyst also showed good ORR catalytic ability. Porous and nanotube-structured Mn_2O_3 catalysts were synthesized and investigated respectively, and the ORR activities of these two catalysts were similar to those of commercial Pt/C [92, 94]. Furthermore, Mn_2O_3 exhibits significantly higher activity than other Mn-based oxides. Manganese oxide is a promising

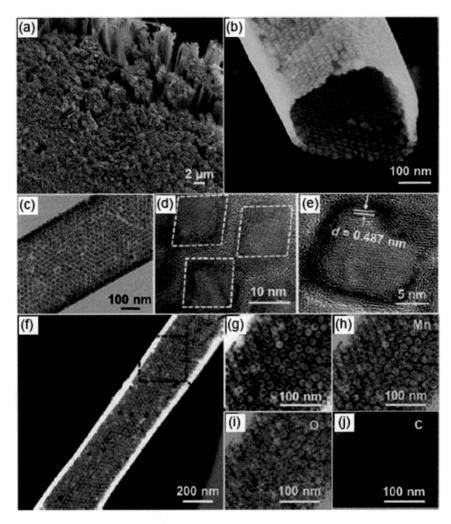

Fig. 2.5 a–d SEM, TEM and HRTEM images of h-Mn₃O₄-TMSLs. e HRTEM image of a single hollow Mn₃O₄ NC. f STEM and g high-magnification STEM images of the region. h–j corresponding elemental mapping images [67]

alternative to noble metals as ORR catalysts due to its low cost, environmental friendliness, and considerable catalytic capacity. However, pristine manganese oxide usually has limited ORR activity in its bulk form, which may be due to its low electrical conductivity and fewer oxygen adsorption sites on the metal oxide surface.

Chen's group developed hydrogenated manganese dioxide (H-MnO$_2$) with nanorod morphology and significantly improved ORR performance through a simple strategy [95]. Compared with pristine manganese dioxide in 0.1 M KOH aqueous solution, the onset potential is positively shifted by about 60 mV and the limiting current density is increased by 14%. Conductivity tests and density functional theory

(DFT) studies indicate that the fast kinetics of ORR is mainly attributed to the improved electronic properties and the enhanced oxygen adsorption capacity is ascribed to the formation of hydrogen bonds. Another convenient and effective surface modification method to enhance the ORR performance of manganese oxide is to combine it with one or more noble metals [96–99], carbon materials [100–102], and other simple metal oxides. Extensive progress has been made based on the achievements of composite catalysts. Carbon materials are one of the best candidates as support materials due to their good electrical conductivity, large surface area, and strong stability. One or more carbon isomers such as porous carbon, activated carbon, carbonitrides, carbon nanotubes (CNTs), carbon nanofibers, and graphene can be used as support materials for ORR. Wang obtained and studied Mn_3O_4/mesoporous carbon composites for the first time [103]. Compared with carbon nanotubes, activated carbon, and graphite, carbon with ordered mesopores (CMK-3) is more suitable to support Mn_3O_4 nanoparticles with diameters of about 10 nm. The composite materials can provide a sufficient effective three-phase interfacial area, which is favorable for the oxygen reduction reaction. The Mn_3O_4 nanoparticles were further coupled with nitrogen-doped graphene by a hydrothermal process and exhibited high ORR activity, good durability, and tolerance to methanol [104–106]. A MnO dual-carbon coating design with high electrocatalytic activity and superior stability was obtained by a two-step coating of Mn_3O_4 followed by heat treatment. Figure 2.6 shows the target catalyst of N-doped carbon-coated MnO nanoparticles in a graphene matrix [107]. Polydopamine and GO were introduced onto the surface of Mn_3O_4 nanocrystals before heat treatment, which was shaped like square prisms. As Mn_3O_4@PDA nanocrystals were encapsulated in graphene, GNS@MnO@N-doped carbon (GMNC) with better ORR catalytic activity was formed. The onset potential of GMNC was 0.98 V (vs. RHE), which was higher than that of Pt/C, while the half-wave potential of CMNC was almost comparable to that of Pt/C. The unique design of the N-doped double carbon coating generates a large number of defects, which are beneficial to the improvement of electrochemical performance and electronic conductivity.

Recently, Ti_3C_2 MXene with a two-dimensional (2D) structure has been used as support material for manganese oxides. MXene exhibits superior electrical conductivity, good chemical stability, and high surface area, all of which favor ORR. Mn_3O_4/MXene nanocomposites were first developed by Xue et al. The favorable electrochemical activity for ORR was in the dominant four-electron transfer process with an onset potential of 0.89 V (same as Pt/C) [108]. The Zn-air battery with Mn_3O_4/MXene as a cathode catalyst gave high open circuit potential (1.37 V) and high power density (150 mW cm^{-2}).

In general, considering its lower price and desirable ORR activity, silver as a noble metal is used to modify MnO_x. The silver-decorated Mn_3O_4/C composites can be synthesized by two different preparation procedures: the citrate protection method and the simple thermal decomposition method [98, 99]. The onset potential of the Ag/Mn_3O_4/C catalyst is about 0.92 V (vs. RHE), which is close to that of Pt/C. The enhancement of its ORR activity is caused by two main effects, namely ligand and strain effects. Ligand effects increase d-band vacancies, both of which increase the

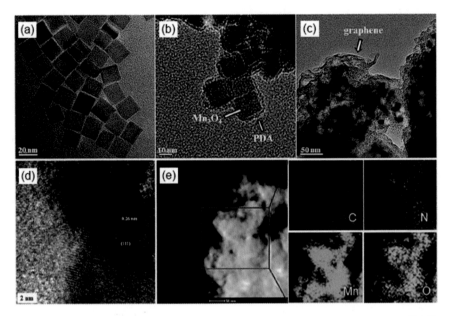

Fig. 2.6 TEM images of **a** Mn_3O_4 nanocrystals, **b** Mn_3O_4@PDA nanocrystals, and **c**, **d** GMNCs at different magnifications. **e** Dark-field TEM image and corresponding EDS elemental map image [107]

kinetics of O-O bond cleavage in favor of ORR activity. The Ag-Mn_3O_4/C catalyst synthesized by the simple thermal decomposition also showed good ORR activity in alkaline medium [99]. The onset potential of Ag-Mn_3O_4/C in 1 M NaOH is ~ 0.11 V (compared to SCE), which is higher than that of Ag/C. The formation of Ag/Mn-O-C bonds facilitates the transfer of covalent electrons from Ag to Mn_3O_4 through the carbon matrix, which played a key role in enhancing the ORR activity of the composites. In addition, Mn_3O_4 could reduce the chemisorption of oxygen reduction intermediates on the surface of Ag particles according to the spillover effect, thereby significantly increasing the electrochemically active surface of Ag.

The incorporation of other low-valent elements, such as Ca, Ag, Mg, and Ni, is also an effective way to enhance the catalytic activity of manganese oxides for ORR [109–113]. Roche et al. found that manganese oxide/C incorporating Ni/Mg could further enhance its ORR specific activity, bringing it closer to the benchmark 10 wt% Pt/C. Even rock-salt-type MnO, which is considered to be one of the worst ORR catalysts in the MnO_x group, can achieve high electrocatalytic ability by incorporating Ca. The ORR activity and stability of Ca-substituted MnO/C ($Ca_{0.5}Mn_{0.5}O$/C) were comparable to those of commercial Pt/C in alkaline media. The greatly enhanced catalytic activity of $Ca_{0.5}Mn_{0.5}O$/C is mainly attributed to changes in surface chemistry, including the presence of divalent redox pairs of Mn^{2+}/Mn^{3+} and the formation of MnOOH.

2.3.1.2 Simple Cobalt-Based Oxide Catalysts

Single cobalt oxides have been extensively studied as efficient bifunctional catalysts. Compared with other metal oxides, they have an excellent catalytic ability for both ORR and OER. Here, the ORR properties of single cobalt oxides are mainly reviewed.

Cobalt oxides, such as CoO and Co_3O_4, have good ORR catalytic ability. Spinel-type Co_3O_4, with tetrahedral and octahedral sites occupied by Co^{2+} and Co^{3+}, is the most commonly used ORR catalyst. The surface structure of Co_3O_4 also affects its ORR catalytic ability. Due to the above reasons, morphology control and surface modification are effective strategies to enhance the catalytic ability of ORR. Co_3O_4 can be tuned into different morphologies, such as core/shell [114], ordered mesopores [115], nanosheets [116, 117], nanotubes [118], nanoflowers [119], nanomembranes [120, 121], nanospheres [122], nanospheres [123], these all show good ORR ability. Mesoporous nanowire arrays of nitrogen-doped Co_3O_4 were fabricated and applied to a flexible solid-state zinc-air battery via a facile hydrothermal method in NH_4F solution [124]. Fig. 2.7a–c show Co_3O_4 nanowire clusters. The ordered N-doped nanowire arrays remained unchanged after heat treatment under NH_3 atmosphere. Uniform N dopants can introduce crystal defects, possibly caused by stacking faults, which are confirmed in Fig. 2.7d–f. The ORR activity of N-doped Co_3O_4 nanowires grown on carbon cloth is much higher than that of Co_3O_4 and most Co-based catalysts/carbon materials with an onset potential of 0.94 V (vs. RHE). DFT calculations were also used to reveal the effect of N dopants on the ORR activity of Co_3O_4, and it was found that doped N could enhance the electrical conductivity and O^{2-} adsorption.

To further enhance the catalytic ability of Co_3O_4, combining it with carbon materials was identified as an effective approach [125, 126]. Carbon materials, including graphene, carbon nanotubes, and porous carbon, generally possess high electrical conductivity and large surface area, and are recognized as promising catalyst substrates. Combining cobalt oxides with carbon materials and constructing suitable nanostructures can significantly improve catalytic activity by providing increased catalytic surface area, high electrical conductivity, and modified surface conditions. Compared with Mn_3O_4 and Co_3O_4 on GO, the graphene-supported Co_3O_4-based composite (Co_3O_4-Mn_3O_4/GO) exhibited higher electrocatalytic activity and durability for ORR [127]. Spinel Mn_3O_4 grows preferentially along the (111) plane of the Co_3O_4 particles supported on graphene, forming a unique three-dimensional oxide-oxide stacking structure. XPS confirmed the strong interaction of Co and Mn, leading to the covalent electron transfer between Co_3O_4 and Mn_3O_4, thereby enhancing the ORR activity of the composite catalyst, as depicted in Fig. 2.8. Combining with noble metal catalysts and doping with other metal elements are also feasible and effective strategies to improve the electrocatalytic performance of cobalt oxides [128].

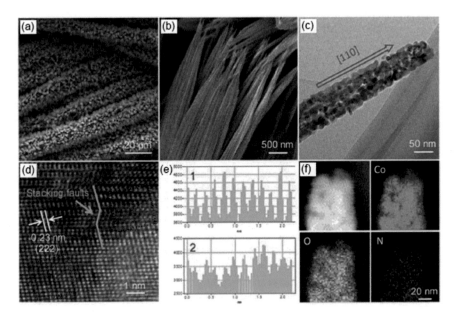

Fig. 2.7 SEM images of **a** Co_3O_4 and **b** N_2-Co_3O_4. **c, d** TEM images of N_2-Co_3O_4. **e** Contrast intensity curves of the two squares in (**d**). **f** High-angle annular dark-field STEM image and corresponding elemental map of N_2-Co_3O_4 [124]

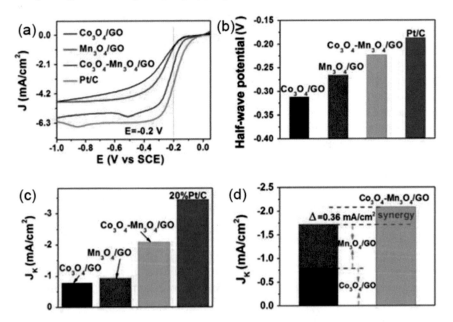

Fig. 2.8 **a** Comparison of ORR polarization curves in O_2-saturated 0.1 M KOH solution with a rotation speed of 1600 rpm and a sweep speed of 5 mV s^{-1}. **b** Half-wave potentials of all samples; **c, d** Kinetic current densities (J_k) of all samples at -0.2 V (compared to SCE) [128]

2.3.1.3 Perovskite Catalyst

Besides oxides of transition metals manganese and cobalt, other simple oxides of transition metals such as Ni [129, 130], La [131–133], Fe [134–138], and Ce [139, 140], have also been shown to be active for ORR. However, the catalytic activity of these oxides is generally not as good as the manganese and cobalt oxides mentioned above. Incorporation with functionalized carbon materials or other metal oxides proved to be a convenient way to promote the ORR catalytic activity of such simple transition oxides. A great deal of work has been done on the modification of transition metal oxides and they can be found in other review articles [141–143].

The general formula for perovskite oxides is ABO_3 (A = rare earth metal or alkaline earth, B = transition metal). Various perovskites can be obtained by partially replacing the cations at the A or B sites, with the formula $A_{1-x}A'_xB_{1-y}B'_yO_3$. The perovskite structure can be roughly estimated according to the tolerance factor (t). For perovskite materials, it is defined as $t = (r_A + r_O)/sqrt(2)(r_B + r_O)$, where r_A, r_B, and r_O represent the ionic radii of the A-site, B-site, and oxygen elements, respectively. Over the past few decades, perovskite materials have attracted extensive interest as an important ORR catalyst in alkaline solution due to their compositional and structural peculiarities. Research on perovskites for ORR dates back to the 1970s [144], when Matsumoto et al. reported $LaNiO_3$ perovskite materials for oxygen electrocatalysis [145, 146]. Factors that determine the catalytic activity of perovskite ORR catalysts include intrinsic activity, phase structure, morphology, specific surface area, electronic conductivity, surface properties, oxygen uptake behavior, elemental vacancies, and types of conductive additives (Fig. 2.9a).

2.3.2 Transition Metal Nitride/Sulfide/Phosphides

Transition metal nitrides with high electrical conductivity and good corrosion resistance exhibit higher ORR activity and stability than transition metal oxides. For example, the tubular combination of $Ti_{0.8}Co_{0.2}N$ nanosheets whose electronic structure is modified by Co doping exhibits appreciable ORR performance. Other nanoparticles, ZrN, possess strong d-orbital states, stable Zr end faces with similar oxygen adsorption energies, and high electron localization on small Zr-O clusters, showing excellent ORR activity and stability, even exceeding the alkaline environment (Fig. 2.9b) [147]. Likewise, integrating carbon materials with nitrides can also improve ORR performance. For example, a novel favorable ORR electrocatalyst was synthesized by mounting $TiCoN_x$ nanoparticles on N-doped reduced graphene oxide supports [148]. The synergy between $TiCoN_x$ and the support and the apparent increase in active pyridine N is responsible for overtaking the ORR activity.

Similar to oxides and nitrides, metal sulfides also exhibit some ORR catalytic activity. Interface engineering, defect engineering, morphology, and electronic structure tuning are also good ways to optimize their intrinsic activity. With the application of nanocrystal interface engineering at the atomic level, CuS/NiS_2 with well-defined

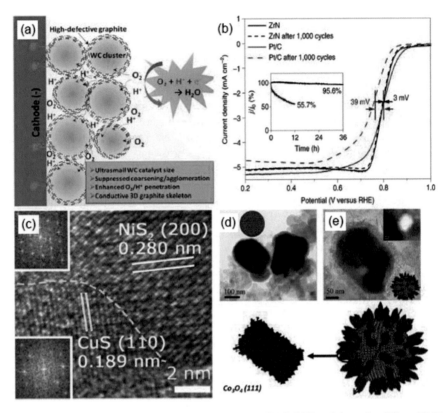

Fig. 2.9 **a** Schematic diagram of ORR catalytic process [67]. **b** ORR activity and stability of ZrN and Pt/C in O^{2-} saturated 0.1 M KOH solution [147]. **c** HRTEM image of the CuS/NiS$_2$ interface nanocrystal and corresponding FFT pattern (inset) [149]. HRTEM images of Co$_2$P-40 **d** before ORR and **e** after ORR, and the corresponding schematics are drawn below [154]

defects and interfaces exhibits superior ORR catalytic performance compared to commercial Pt/C (Fig. 2.9c) [149]. As another example, the electronic structure of the active site in Co$_3$FeS$_{1.5}$(OH)$_6$ was tuned by anions, showing high ORR activity comparable to that of Pt/C [150]. Furthermore, the Co$_9$S$_8$/C nanosheets possessed abundant active sites and nanosheet structures, which largely contributed to their excellent ORR catalytic performance and enormous long-term stability [151]. Combining phosphorus with transition metals could also increase the stability and corrosion resistance of the transition metals themselves, thus offering great potential for ORR electrocatalysts [152, 153]. However, metal phosphides with poor electronic conductivity hinder their electrocatalytic activity and electron transfer process. Combining metal phosphides with carbon materials could improve their surface area and electrical conductivity. However, during the ORR process, the dissolution and precipitation of cobalt oxides on the surface of Co$_2$P particles were involved in their excellent stability in acidic and alkaline environments (Fig. 2.9d, e) [154]. Other chromium phosphides also showed the advantages of high thermal stability, good oxidation resistance, and corrosion resistance. When mixed with carbon, the strong oxygen adsorption on the surface of CrP promoted the splitting of O_2, showing excellent ORR activity and stability in alkaline environment.

Transition metal phosphides (TMPs) are traditionally obtained by metal phosphating in solution-thermal or solid-state thermal methods. Phosphating agents such as red/black phosphorus, NaH_2PO_2, trioctyl phosphorus, etc., are traditionally used in solid-state processes, which involve the generation of toxic phosphating gases. Phosphating of transition metals using phytic acid (PA) as an alternative phosphating agent has been known since 2015. In the past, various metal phosphides, such as FeP, CoP, MoP, etc., have been synthesized using PA for electrochemical water separation, oxygen reduction, etc. [155–157]. Although the electrocatalytic activity of TMPs for water separation has been widely studied, the research on oxygen electrocatalysis is limited. Sing et al. are the first to study the ORR activity of metal phosphide-based electrocatalysts, although their activities are inferior to conventional catalysts. Recently, a new method for phosphating Co without using any conventional phosphating agents was demonstrated [158]. Carbothermal reduction-assisted phosphating of metal precursors (cobalt/iron polypyridyl complexes) was achieved at high temperatures under an inert atmosphere. The PF_6-anion of the complex was used for the phosphating of metals for the first time. The amount of carbon in the precursor complex controls the reduction and subsequent phosphating of the precursor. For example, thermal annealing of the [Co-(pyterpy)$_2$](PF$_6$)$_2$ composite yielded nitrogen-doped carbon-coated CoP(CoP@NC). However, polypyridyl complexes such as [Co(bpy)$_3$](PF$_6$)$_2$ and [Co(terpy)$_2$](PF$_6$)$_2$ did not yield ideal CoPs because of the lack of the required carbothermic reduction process (Fig. 2.10). Thermal annealing of these complexes with an additional carbon source such as sucrose or melamine yielded the desired CoP encapsulated/intercalated in carbon. The CoP@NC nanostructures possessed high ORR activity, favoring the ideal four-electron reduction of oxygen to water, and were very durable. The synergistic effect between the core CoP and the encapsulated nitrogen-doped carbon enhanced the overall performance. Among various TMPs, iron phosphide has received special attention because Fe-P bonds could activate oxygen molecules and facilitate electron transfer kinetics. Recently, the electrocatalytic activity of nitrogen- and phosphorus-doped carbon-supported Fe_2P nanostructures and their possible application to PZAB have been investigated [159, 160]. These phosphate-based PZABs have achieved specific capacities of 577–691 m Ah g^{-1} and energy densities of 672–861 Wh kg$_{Zn}^{-1}$.

The ORR activity of transition metal nitrides has been known since 1966 [161]. However, nitride-based catalysts have not attracted much attention for many years. Recent reports suggest increased interest in transition metal nitride-based catalysts, possibly due to the increasing demand for non-palladium electrocatalysts for energy conversion applications [162]. Among various transition metal nitrides/carbonitrides, TiN, TiCN, MoN, Ni_3N, CoN, Cu_3N, etc. have received extensive attention [163–165]. Wu and Chen synthesized Cu_3N nanocubes in a single-phase process in organic solvents and investigated their ORR activity at alkaline pH. Subsequently, the ORR activity of Cu_3N supported by nitrogen-doped carbon nanotubes obtained by atomic

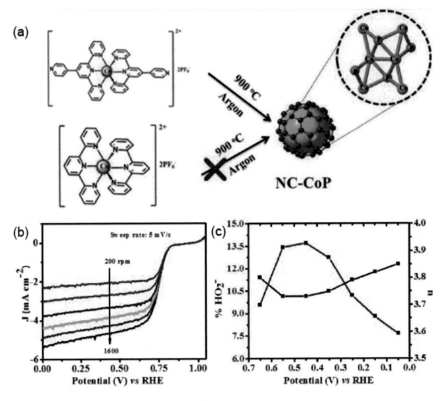

Fig. 2.10 **a** Synthesis schematic of NC-CoP. **b** ORR polarization curves and **c** hydrogen peroxide yield and electron transfer number of NC-CoP catalysts [153]

layer deposition was investigated and the total activity of 118.5 mA mg^{-1} was achieved. However, in-depth studies on the persistence and dynamics of ORR are still lacking [166]. The N-rGO/Cu$_3$N electrocatalyst showed good ORR electrocatalytic activity with an onset potential of 0.92 V and a mass activity of 163 mA mg^{-1}, which was significantly higher than that of CNT-supported Cu$_3$N. The catalyst supported by N-rGO greatly contributed to the ORR activity and durability of the catalyst. Abruna's group demonstrated the enhanced ORR performance of cobalt nitride-based electrocatalysts [167]. Compared with the carbon-supported Co$_3$O$_4$ catalyst, the mass activity of the Co$_4$N/C catalyst was increased by 8 times at a potential of 0.85 V. The enhanced activity was attributed to a thin layer of conductive nitride core and oxide shell. Li et al. demonstrated the PZAB performance of isolated cobalt nitride and vanadium nanostructures encapsulated with nitrogen-doped carbon nanosheets and achieved a peak power density of 237.8 mW cm^{-2} at 10 mA cm^{-2}. It provided a nearly stable discharge voltage for 24 h under the discharge current density. A recent theoretical study predicted that 2D RuN$_2$ could catalyze ORR at a more positive onset potential than Pt(111) electrodes.

In general, transition metal nitrides (TMNs) are known to be more stable compared to oxides because of the high triple bond energy between metal and N atoms [168, 169]. In fact, TMNs have long been investigated as promising ORR electrocatalysts, given their high electrical conductivity and good corrosion resistance. Unfortunately, their ORR activity in acidic electrolytes is suboptimal even at the RDE level. However, it has been reported that binary titanium nitride synthesized by incorporating Ni in TiN nanocrystals showed high ORR activity in acidic electrolytes, which was almost comparable to that of commercial Pt/C catalysts in alkaline electrolytes [170]. It has been suggested that the incorporation of Ni might enhance the ability of Ti atoms to transfer electrons to adsorbed oxygen molecules while reducing the Ti–O strength to a moderate level, thereby contributing to the enhanced ORR activity. By the same doping method, the VN- and NbN-based materials also showed good ORR activity in alkaline media. With a deeper understanding of the structure–activity relationship, $Ti_{0.8}Co_{0.2}$ N nanotubes constructed from thin $Ti_{0.8}Co_{0.2}$ N nanosheets exhibited extremely high performance compared to their NP-patterned counterparts at the full-cell level of ORR activity [171]. Similar to transition metal oxides, the hybridization of TMN with carbonaceous supports could also benefit ORR performance. For example, a hybrid ORR catalyst composed of TiCoNNPs and N-doped reduced graphene oxide (denoted as TiCoN/N-rGO) exhibited excellent ORR activity in alkaline media. It has been proposed that the synergistic effect resulting from the strong interaction between TiCoN and N-rGO in the hybrid material contributed to the excellent ORR activity. Similar to transition metal oxides, the hybridization of TMN with carbonaceous supports will also benefit ORR performance. For example, a hybrid ORR catalyst composed of TiCoNNPs and N-doped reduced graphene oxide (denoted as TiCoN/N-rGO) exhibited excellent ORR activity in alkaline media. It has been proposed that the synergistic effect resulting from the strong interaction between TiCoN and N-rGO in the hybrid material contributed to the excellent ORR activity.

Carbides of Groups IV to VI metals have shown to display platinum-like catalytic properties in many reactions, and their desirable ORR activity has also been demonstrated [172–174]. However, unlike TMNs, transition metal carbides (TMCs) cannot tolerate acidic conditions, so TMCs are usually encapsulated in carbon and/or graphitic layers. Guo et al. reported the synthesis of WC NP catalysts with an average diameter of only 19 ± 0.9 nm, which were fully encapsulated in graphitic layers (WC@C) [175]. Xiao et al. proposed a facile pyrolysis method to prepare iron carbide (Fe_3C) encapsulated in nitrogen-doped graphite layers (Fe_3C/NG), and the resulting catalyst exhibited excellent ORR activity in both acidic and alkaline solutions [176]. However, the performance of TMC-carbon composite catalysts may be affected by many factors, such as the N content and types in the graphite shell, since N-doped carbon may also be highly active for ORR. Furthermore, the exact composition in TMC can be very complex, for example, tungsten carbide might be a mixture of WC and W_2C, while an iron carbide sample might include Fe and Fe_3C particles. Therefore, more theoretical and in-depth investigations are necessary to fully understand the role of individual components, and the activity-enhancing mechanism of TMC-based ORR electrocatalysts.

Transition metal phosphides and sulfides have also been considered potential ORR electrocatalysts because of their abundant valences of P and S atoms, which may lead to different redox couples in the catalytic process. For example, both CoP and FeP showed good ORR activity and persistence in acidic media due to the inertness of metal phosphides. However, the ORR performance of these catalysts should be further improved as it was still much worse compared to the Pt/C catalysts. In another work, N-doped Co_9S_8/graphene composites with abundant defects also showed rather high activity and persistence in alkaline media [177]. It has been proposed that NH_3 plasma treatment of Co_9S_8/graphene hybrids can achieve N doping and etching into Co_9S_8 and graphene. The introduction of N will tune the electronic properties of Co_9S_8 and graphene, while the surface etching will make more active sites available, all of which greatly contribute to the ORR activity.

2.3.3 Transition Metal Single Atoms

Transition metal single-atoms are catalysts in which metal active sites are dispersed on atomic-scale support. It has recently emerged as a hot research front in the field of ORR electrocatalysts [52]. With atom utilization up to 100% and intriguing properties, SACs are considered to have great potential in reducing the use of PGM. Based on a brief review of the SACs literature, it is not difficult to see that SACs are not simple additions to the ORR electrocatalyst family (i.e. PGM, non-PGM, and carbon-based materials), but can be seen as an important conceptual advance. The performance of all types of catalysts can be radically improved. More specifically, PGM-containing SACs exhibited comparable or superior ECSAs to their counterparts in other patterns, such as NPs or clusters, even with much lower PGM loadings [178]. Improving the utilization efficiency of PGM is an ideal strategy to reduce the consumption of noble metals without sacrificing catalytic performance, which has important implications for reducing catalyst cost and commercial implementation. Furthermore, the demetallization effect that plagues carbon-based materials may be overcome by SACs. Although many research groups have obtained high ORR performance on different SACs, a more comprehensive investigation is now urgently needed, which should not only provide detailed mechanistic studies of ORR performance and RDE levels but also, more importantly evaluate MEA tests or metals-Air batteries systemically.

In the early 1970s, macrocyclic compounds containing $M-N_4$ (M=Fe and Co) molecules, such as porphyrins and phthalocyanines, were first identified as active sites for ORR. The metal center dependence of the ORR activity of this molecular catalyst could be explained by molecular orbital (MO) theory, showing a trend of Fe > Co > Ni > Cu $\cong$ Mn. This finding is instructive to understand the catalytic mechanism of M-N-C SAECs. Significant attention has been paid in recent years to the development of PGM-free electrocatalysts for ORR.

2.3.3.1 Iron-Based Single-Atom Catalysts

DFT calculations indicate that Fe and N co-doped graphene can form FeN_4 structures, which are as active as Pt metals in ORR. Compared with other precursors, ZIF-8 has attracted much attention as a sacrificial template for the synthesis of Fe-N-C SAECs because the unique microporous structure is favorable for the accommodation of Fe and N to obtain a high-density FeN_4 site after simple high-temperature pyrolysis. Despite the emergence of pyrolysis-free strategies, thermopyrolysis of composite precursors containing Fe, N, and C remains the dominant method for the synthesis of Fe-N-C SAECs.

Lai et al. demonstrated a host–guest strategy to construct Fe nanoclusters (guest)@ZIF-8 (host) precursors to obtain Fe-N-C SAECs. The uniform dispersion of Fe species in the ZIF-8 precursor leads to a uniform distribution of Fe single sites in the derived electrocatalyst. The obtained Fe-N/C catalyst with abundant Fe-N_4 patterns provided a current density of 5.12 mA cm^{-2} at 0.4 V_{RHE}. Chen et al. designed a synthetic route starting from iron-doped ZIF-8 precursors (Fe-ISAs/CN, Fig. 2.11a) to obtain isolated iron atoms anchored on N-doped porous carbon [179]. The Fe-N_4-based catalyst showed comparable ORR activity to commercial platinum catalysts in 0.1 M KOH after the removal of heat-treated by-product iron nanoparticles (Fig. 2.11b). Wu and coworkers constructed Fe-N_4 patterned Fe SAECs by exchanging Zn cations in ZIF-8 precursors with iron before thermal activation (Fig. 2.11c) [180]. Fe SAEC obtained from 1.5 at% Fe-doped ZIF-8 precursors (1.5 Fe-ZIF) were found to be most active for ORR at $E_{1/2}$ of 0.85 V_{RHE} in O_2 saturated electrolyte (Fig. 2.11d). They believe that the key is the high yield of a single iron site. Fe-N_4 with low site density was generated due to reducing the iron loading in the precursor, while higher loadings led to the formation of inactive agglomerates or nanoparticles. In addition to pyrolyzing the seemingly expensive imidazoline frameworks, researchers have also been looking for synthetic methods that utilize fewer or less expensive ligands. Chen et al. reported the pyrolysis of Fe^{3+} and SCN^- decorated CNTs [181]. Two main peaks were found at 164.0 and 165.1 eV in the XPS spectrum of S 2p electrons, which were attributed to the sulfur dopant (-C-S-C-) (Fig. 2.11e). The introduction of N and S dopants induced an inhomogeneous charge distribution in the carbon framework, resulting in positively charged carbon atoms, which were thought to enhance the adsorption of oxygen species to Fe-N_x sites (Fig. 2.11f). Polymer hydrogels were also used as 3D carbon/nitrogen precursors to derive Fe SAECs.

Zhang et al. systematically investigated the effect of precursor particle size and thermal activation conditions on the ORR performance of Fe-N-CSAECs [182]. Fe-doped ZIFs were synthesized with controlled particle size (Fig. 2.12a, b). STEM and EXAFS studies showed that iron was atomically dispersed through coordination with N and C in the carbon matrix (Fig. 2.12c, d). As shown in Fig. 2.12e, the best SAEC came from Fe-ZIF with a particle size of 50 nm, achieving an $E_{1/2}$ of 0.85 V for RHE, which was only 30 mV lower than that of Pt/C. The ORR activity was found to decrease with increasing Fe-ZIF precursor particle size (Fig. 2.12f). The

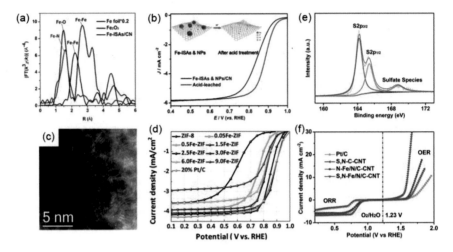

Fig. 2.11 **a** Fourier transform (FT) of Fe K-edges and wavelet transform (WT) of Fe-ISAs/CN. **b** ORR polarization curve in O_2 saturated 0.1 M KOH [179]. **c** The same HAADF-STEM image depicts single Fe dots (bright spots) dispersed throughout the carbon phase of the 1.5Fe-ZIF catalyst. **d** RRDE ORR polarization map in O_2 saturated electrolyte [180]. **e** XPS spectrum of S 2p of S, N-Fe/N/C-CNT. **f** Overall polarization curves of the entire ORR and OER regions of all samples in 0.1 M KOH solution. The oxygen electrode activity (ΔE) was evaluated by the OER potential at a current density of 10 mA cm^{-2} and the half-wave potential at a current density of -3 mA cm^{-2} [181]

effect of pyrolysis temperature on the Fe-N-C electrocatalyst was further investigated. The electrocatalyst was derived from 50 nm precursor nanoparticles. It was found that 800 °C was the lowest temperature to activate the formation of Fe-N-C sites for ORR activity (Fig. 2.12g). The ORR activity increased with the pyrolysis temperature up to 1100 °C (Fig. 2.12h). This is due to the loss of pyridine and fivefold bonded pyrrolidine and the increase of graphitic at higher temperatures. Pyridine-like N provides coordination sites for atomic iron in the form of FeN_x, while graphitic-like N affects the geometric and electronic structures of carbon substrates. Wu et al. investigated the formation mechanism of FeN_4 site by anchoring Fe cations on nitrogen-doped carbon followed by controlled thermal activation [183]. Unlike previous understanding, thermal activation only required a relatively low temperature (400 °C) to form active FeN_4 sites, although 700 °C gave the best ORR performance. Further increasing the pyrolysis temperature tended to produce the loss of N dopants, resulting in a decrease in inactivity.

It is found that coating ZIF with polymer or inorganic thin films can introduce mesopores into the derivatized SAECs, increase atom utilization efficiency, and also improve mass transport [184]. It is found that coating ZIF with polymer or inorganic thin films can introduce mesopores into the derivatized SAECs, which increases atom utilization efficiency and improves mass transport. Zhang et al. constructed concave Fe-N-C SAECs with mesoporous and extended surface area with mesoporous silica

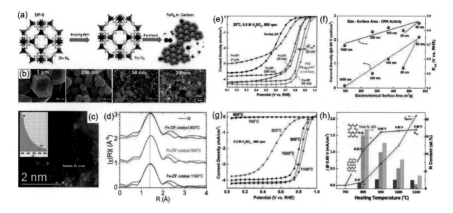

Fig. 2.12 **a** Synthesis of iron-doped ZIF-derived catalysts. **b** The size of the precisely controlled Fe-ZIF catalyst. **c** HAADF-STEM image and EELS analysis of the iron-doped ZIF catalyst [inset of panel (**c**)]. **d** EXAFS spectra and fittings of catalysts were prepared at different temperatures. **e** ORR polarization maps of Fe-ZIF catalysts with different particle sizes. **f** Correlation between ORR activity and particle size. **g** ORR activities of Fe-ZIF catalysts prepared at different temperatures. **h** Correlation between ORR activity and heating temperature-dependent new bond formation [182]

(mSiO$_2$) coating strategy (Fig. 2.13a) [185]. The non-uniform shrinkage of ZIF-8 during heat treatment is caused by the mSiO$_2$ thin film, resulting in the outer depressions of the nanoparticles and abundant mesopores inside. Highly dense atomically dispersed iron atoms existed in the mesoporous carbon polyhedra as shown by HAADF-STEM measurements (Fig. 2.13b). The structure of a possible FeN$_4$C$_8$ molecule has been proposed with two O$_2$ molecules adsorbed at the active site in an end-pair mode (Fig. 2.13c). Figure 2.13d showed that the E$_{1/2}$ gap of TPI@Z8(SiO$_2$)-650-C and TPI@Z8-650-C was only 12 mV. Nonetheless, the kinetic activity differences in fuel cells are significantly amplified due to the high catalyst loading in the membrane electrode assembly (MEA). Under the DOE test protocol, TPI@Z8(SiO$_2$)-650-C achieved current densities of 0.022 A cm^{-2} (@0.9 ViR-free) and 0.047 A cm^{-2} (@0.88 ViR-free), which exceeded DOE's 2018 target (Fig. 2.13e). As shown in Fig. 2.13f, TPI@Z8(SiO$_2$)-650-C possessed a high density of FeN$_4$ molecules and a large surface area, which provided a large number of active sites at the three-phase boundary (TPB) of the fuel cell. Efficient mass transport in the catalyst layer is expected to facilitate most of the TPB active sites, enabling them to participate in ORR even in the high current region.

2.3.3.2 Cobalt-Based Single-Atom Catalysts

Although significant progress has been made in promoting the activity of ferrous SAECs, the issue of their durability remains to be resolved. Typically, radical species generated by the iron-catalyzed Fenton reaction will result in a loss of activity for

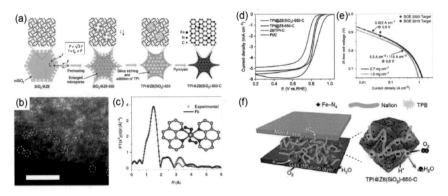

Fig. 2.13 a Schematic diagram of the synthesis process of TPI@Z8(SiO$_2$)-650-C. **b** HADDF-STEM image. Scale bar: 2 nm. **c** FT-EXAFS profile (inset: predicted structure of FeN$_4$C$_8$ molecule with two O$_2$ molecules adsorbed in end-to-end mode). The brown, blue, grey and red spheres represent iron, nitrogen, carbon, and oxygen atoms, respectively. **d** ORR performance in O$_2$ saturated electrolyte. **e** Tafel plots of the activity at 0.9 V and the bulk activity at 0.8 V of TPI@Z8(SiO$_2$)-650-C. **f** Schematic illustration of three-phase boundary (TPB) active sites in three representative Fe-N-C catalyst-based electrodes [185]

PGM-free catalysts. To mitigate the Fenton reaction induced by iron, i.e., the generation of radical species that attack catalysts and ions, there is an urgent need to develop PGM-free and iron-free catalysts for low cost and robustness. Yin et al. developed stable Co SAECs by pyrolyzing predesigned Zn/Co bimetallic MOFs, and the introduction of Zn can effectively prevent Co atoms from agglomerating and generating micropores during pyrolysis. The obtained Co-N$_x$ single point exhibited remarkable ORR performance (E$_{1/2}$ = 0.881 V) and durability in alkaline electrolytes. Recently, catalysts with atomically dispersed CoN$_4$ sites were synthesized by a simple one-step thermal activation of Co-doped ZIF precursors [186]. The best Co SAECs obtained by optimized thermal activation conditions possessed excellent ORR activity (E$_{1/2}$ = 0.80 V) and good stability in acidic media. However, severe aggregation occurred at high cobalt loadings, which limited the maximum ORR activity of this type of catalyst.

An innovative surfactant-assisted MOFs approach was later reported to construct core–shell Co SAEC@F127 with a significantly increased density of single-atom Co-active sites [187]. This is inspired by the strong interaction between the surfactant F127 block copolymer layer and the ZIF precursor particles (Fig. 2.14a). The carbon shell derived from the F127 layer effectively retained the dominant micropores and high content of N in the carbon matrix during pyrolysis, enabling the formation of highly dense atomically dispersed Co-active sites (Fig. 2.14b). The configuration of the CoN$_{2+2}$ active site was proposed by fitting a Fourier transform of the k^2-weighted EXAFS data (Fig. 2.14c). In the acidic electrolyte, the ORR activity of the Co SAEC@F127 catalyst was significantly enhanced with an E$_{1/2}$ of 0.84 VRHE, which was comparable to the state-of-the-art Fe SAEC catalyst (Fig. 2.14d). Figure 2.14e depicted the transition states and predicted the activation energies for

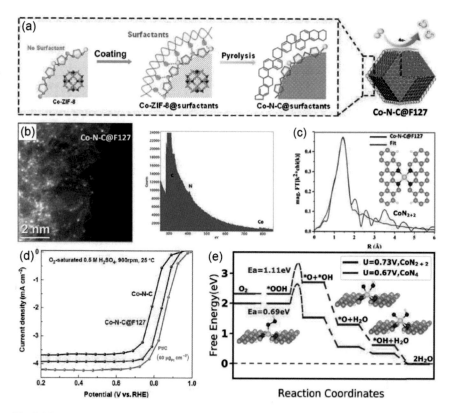

Fig. 2.14 **a** In situ closed pyrolysis strategy to synthesize core–shell Co-N-C@F127. **b** AC-HAADF-STEM image and accompanying EELS dot spectrum of Co-N-C@F127. **c** Co K-edge FT-EXAFS profile and fitted curve of the Co-N-C@F127 catalyst (inset: CoN$_{2+2}$ structure). Grey, blue, yellow, and white balls represent C, N, Co, and H atoms, respectively. **d** ORR polarization map. **e** Calculated free energy evolution of the 4e$^-$ ORR pathway at the CoN$_{2+2}$ site and the CoN4 site at the limiting electrode potential. Atomic structures of the initial state (left), transition state (middle), and final state (right) of the OOH dissociation reaction at the CoN$_{2+2}$ site [187]

the dissociation of *OOH on CoN$_{2+2}$ and CoN$_4$ structures. The Co SAEC@F127 catalyst generated a large number of CoN$_{2+2}$ active sites, which could effectively catalyze ORR through the 4e$^-$ pathway.

2.3.3.3 Manganese-Based Single-Atom Catalysts

Despite extensive work on Fe-N-C and Co-N-C electrocatalysts, their durability for ORR is still insufficient. Especially under ideal high-potential reaction conditions (e.g., > 0.6 V vs. RHE), this limits their practical application in PEMFCs. Preliminary DFT calculations suggest that manganese-based SAECs containing MnN$_4$ molecules intercalated in carbon may have comparable activity but greater stability relative to iron-based SAECs. Therefore, manganese-based SAECs have received increasing attention in developing PGM-free electrocatalysts for ORR.

Xiong et al. reported the deposition of atomically dispersed manganese-nitrogen-carbon films (2.5 ± 0.2 nm thick) (denoted as f-MnNC/CNTs in their report) on oxygen-pretreated CNT substrates (Fig. 2.15a–d) [188]. Mn-N_4 coordination was revealed from K-edge EXAFS analysis (Fig. 2.15e). As shown in Fig. 2.15f, the $E_{1/2}$ of f-MnNC/CNT-170 was 0.83 V (vs. RHE), which was close to that of commercial Pt/C (0.85 V). Bai et al. obtained Mn@NG SAEC by blending graphene with Mn and N through high-temperature pyrolysis and subsequent acid leaching [189]. The EXAFS study of Mn@NG revealed that the main scattering was attributed to the Mn-N_4-C pattern. The $E_{1/2}$ of this catalyst was 0.82 V (Fig. 2.15k). Yang et al. introduced atomic Mn into nitrogen- and oxygen-decorated three-dimensional graphene framework (Fig. 2.15l) [190]. As shown in Fig. 2.15m, a large amount of Mn was atomically dispersed in the carbon matrix. This Mn/C-NO SAEC exhibited an $E_{1/2}$ of 0.86 V, which was even higher than that of Pt/C as a control in this study (Fig. 2.15n). Theoretical calculations showed that the intrinsic catalytic activity of atomic Mn was significantly improved by changing the local geometry of the coordination of atomic Mn, O, and N atoms. Computational results indicated that Mn-N_1O_3, Mn-N_2O_2, and Mn-N_3O_1 were possible ORR active sites (Fig. 2.15o). Compared with Fe- and Co-based SAECs, increasing the density of Mn-N_4 active sites in Mn-N-C electrocatalysts is more challenging because Mn is more likely to form unstable and inactive metal oxides during pyrolysis or carbide. The traditional one-step chemical doping can only introduce a low density of atomic Mn sites. To effectively increase the density of Mn-N_4 active sites in the catalyst, a two-step synthesis strategy involving doping and adsorption processes were reported to obtain the Mn SAECs rich in single-atom Mn sites [191]. In the first step of the synthesis, Mn-doped ZIF-8 precursors were prepared through the conventional chemical doping method. After carbonization and acid leaching, the derived porous carbon was employed as a host to adsorb additional Mn and N sources, followed by thermal activation. This atomically dispersed Mn SAEC achieved promising activity with an $E_{1/2}$ of 0.80 V_{RHE} and excellent stability. By fitting the Fourier transformed EXAFS, an MnN_4C_{12} site was proposed as the active site. DFT calculations further confirmed that the MnN_4C_{12} site displayed an appropriate binding strength with the ORR species and could be the active site for the $4e^-$ pathway. Therefore, the reported Mn SAEC demonstrated an alternative concept to develop a robust and highly active PGM-free catalyst to replace the Fe counterparts.

2.4 Carbon-Based Catalysts

Carbon-based nanomaterials have been widely used as ORR catalysts due to their abundant reserves, low cost, good electrical conductivity, and strong acid and alkali resistance. However, pristine carbon suffers from poor adsorption/activation of oxygen and its intermediates. Recently, modulation strategies including chemical

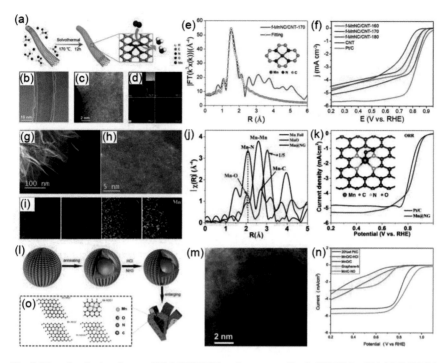

Fig. 2.15 **a** Synthesis scheme of f-MnNC/CNT, and corresponding **b** HRTEM, **c** HAADF-STEM, and **d** elemental mapping images. **e** Mn K-edge FT-EXAFS profile with the proposed structure is shown in the inset. **f** RDE polarization curves in O_2 saturated 0.1 M KOH [188]. **g, h** HAADF-STEM images of Mn@NG at different magnifications. **i** The corresponding EDS elemental maps of C, O, N and Mn. **j** K-edge FT-EXAFS spectrum of Mn. **k** Polarization curves of Pt/C and Mn@NG in O_2 saturated 0.1 M KOH [189]. **l** Schematic diagram of the synthesis process of Mn/C-NO. **m** AC-HAADF-STEM image of Mn/C-NO. **n** RDE polarization curves in O_2 saturated 0.1 M KOH at 10 mV s^{-1} and a scan rate of 1600 rpm. **o** Possible ORR active sites in Mn/C-NO [190]

doping, physical intermolecular charge transfer, and structural defects have been used to develop efficient carbon-based ORR catalysts.

A large number of carbon-based materials have been reported as ORR electrocatalysts, including carbon nanotubes (CNTs), graphene, carbon quantum dots, carbon nanoribbons, carbon cages, and carbon from metal–organic frameworks (MOFs). Carbon-based materials can be roughly divided into two categories, transition metal-doped and heteroatom (non-metal) doped carbon materials. Current carbon-based materials have abundant resources and high ORR performance, which make them promising candidates to replace expensive PGM-based catalysts. Carbon-based materials can be roughly divided into two categories, transition metal-doped and heteroatom (non-metal) doped carbon materials.

Carbon-based ORR electrocatalysts are the most frequently studied besides PGM-based catalysts. Now, some carbon-based electrocatalysts exhibit high ORR activity in alkaline media similar to PGM-based catalysts, the mechanism of high ORR activity of carbon-based materials has been revealed by DFT calculations and advanced characterizations. Unlike ligand and strain effects in PGM-based catalysts, the incorporation of heteroatoms in carbon-based materials can induce charge transfer between doping sites and adjacent carbon atoms, which will effectively lead to charge redistribution, thereby improving the electrocatalytic activity of carbon-based materials. For carbon-based electrocatalysts, it is believed that the ORR active sites of transition metal-doped carbon catalysts are $M-N_4$ molecules and the carbon atom of the pyridine ring of non-metal-doped carbon catalysts.

2.4.1   Undoped Carbon-Based Catalysts

The enhancement of ORR activity of doped carbon-based electrocatalysts is often attributed to different dopants, and their contributions to the origin of ORR activity and intrinsic carbon defects are far from explored. Recently, defect-rich, dopant-free carbons with unusual electronic properties have been proposed as efficient ORR electrocatalysts in both basic and acidic environments [192, 193]. For example, dopant-free carbon nanocages were used to illustrate the effect of intrinsic carbon defects on ORR activity [194]. Pentagonal and chevron edges are two typical roles of carbon defects in ORR improvement. However, conventional carbon nanomaterials have multiple defect sites hindering the identification between performance characteristics and specific carbon defects. In this regard, graphene nanoribbons and exposed chevron carbons with relatively single defects show superior ORRs over other carbon atoms (oxidized chevron/basal plane/armchair edge carbons) and N-doped counterparts.

Besides the doping of other elements, topological and edge defect sites also exhibit unusual electronic properties due to the charge polarization of edge carbon atoms. Thus, some edge-rich carbon materials have been shown to exhibit good ORR activity, even in dopant-free catalysts [195]. For example, Shen et al. reported ORR investigation with air-saturated electrolyte droplets located at specific locations on graphite [196]. This study provided a direct method to compare the intrinsic ORR activity between edge and basal carbon atoms. The results verified that the edge carbon atoms were more active than the basal plane carbon atoms. It is well known that defects can also be created by doping, and defect regions with dangling groups or hydrogen saturation are used to tune the local density of electrons, thereby enhancing ORR activity. Therefore, the combination of doping and defect engineering is often reported as an effective method to enhance the ORR activity of carbon-based materials. Wei and his colleagues synthesized an N-doped carbon catalyst with a large number of internal pores by a unique method called "fixation of shape by salt recrystallization" [197]. Meanwhile, gasification in such a confined space creates a large number of internal pores in the final carbon material. Therefore, it is believed that the number of defect edge sites can be greatly increased by this method, which in turn greatly enhances ORR activity.

2.4.2 Hetero-Atom-Doped Carbon-Based Catalysts

Incorporating heteroatoms (including N, B, S, P, O, F, etc.) into carbon materials is an efficient way to build metal-free alternatives. Due to the difference in electronegativity between carbon and heteroatoms, the incorporation of heteroatoms into carbon nanomaterials can induce local charge redistribution of adjacent carbon atoms, leading to changes in the electronic state, thereby enhancing catalytic performance [198]. For example, doped N atoms could transfer part of the positive charge to adjacent C atoms, resulting in higher conductivity and surface hydrophilicity [199]. Another example of sp-N-doped graphene catalysts showed that O adsorption and electron transfer were promoted, which resulted in superior ORR performance under alkaline conditions, while the acidic ORR activity was slightly lower than that of Pt/C (Fig. 2.16a, b) [200]. An edged pentagonal carbon defect can be selectively created by controllable N doping. The main active sites were shown to be pentagonal defects in highly oriented pyrolytic graphite. Compared with pyridine-like N-sites, it showed higher ORR activity in an acidic environment. In addition, the incorporation of B, P, or S into the sp^2 carbon could also optimize its ORR activity. For example, S/N/P/B doped carbon (carbon nanotube (CNT)/graphene) with stable active sites was constructed by a dual doping strategy [201]. This double-doped carbon showed enhanced ORR catalytic activity and stability due to the synergistic effect caused by electronic interactions between different dopants. Another example is the construction of a unique hierarchical structure of N, S-doped graphite sheets containing steric holes (Fig. 2.16c). This unique structure is beneficial for enhancing the surface area and exposing active sites as well as promoting electron/electrolyte transport properties, thus making the ORR activity and stability superior to other carbon materials.

Since the carbon surface is chemically inert, it is necessary to activate the surface to improve ORR. Heteroatom doping is considered to be an efficient activation method and has been extensively studied. Since Dai and colleagues first reported para nitrogen-containing carbon nanotubes (VA-NCNTs) with excellent ORR activity and high 4e selectivity in 2009, active sites in nitrogen-doped carbons have been extensively studied (Fig. 2.17a). It is believed that N atoms with strong electron affinity break the electroneutrality of the carbon matrix, giving adjacent carbons a high positive charge density [199]. These positively charged carbon atoms can become chemisorption sites for O_2, and then the O_2 molecule is effectively reduced to H_2O through the 4e$^-$ process (Fig. 2.17b, c). However, due to the different chemical states of the N atoms, a controversy about which is more effective for ORR, pyridinic N, or graphitic N was arosed. There are three different views on this issue so far. Nakamura and colleagues constructed a series of model catalysts with well-controlled N species doping to study the effect of each type of N on ORR activity [203]. They found that carbon atoms next to pyridine N with Lewis basicity in N-doped carbon materials were the actual ORR active sites rather than carbon atoms next to graphitic N. However, the electrical conductivity of N-doped carbons was not considered in this study, leading to such results supporting the notion that pyridyl N

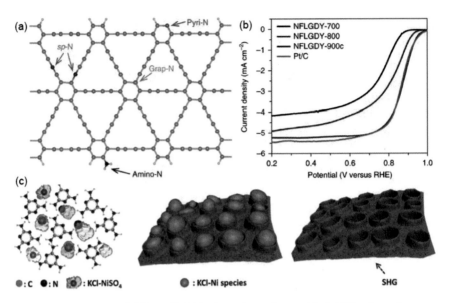

Fig. 2.16 **a** Geometries of different N-doping forms in graphene oxide. **b** ORR polarization curves of various sp-N atom-doped graphene oxides and ORR polarization curves of Pt/C catalysts in 0.1 M KOH at 1600 rpm [200]. **c** Schematic illustration of the fabrication process of N, S-doped graphite sheets [202]

rather than graphitic N is beneficial for ORR activity. In contrast, Dai et al. believed that the quaternary ammonium salt sites were the main reasons for the high activity of galvanically coupled N-doped carbon catalysts for ORR [204]. Likewise, the effects of electrical conductivity and surface N content of N-doped carbon catalysts were ignored in their work. In addition, there is still a third view supported by several other groups, who believe that both pyridine-like N and graphitic N contribute to ORR activity [205, 206]. It is reported that the decisive step for ORR on nitrogen-doped graphene (NG) is the protonation of O_2. When GNG and PNG have the same electrical conductivity, the Gibbs free energy of O protonation is much lower on graphitic N-doped graphene (GNG, 0.921 eV) than on pyridine N-doped graphene (PNG, 2.037 eV) (Fig. 2.17d). The initial ORR activity of GNG was higher than that of PNG. However, the electrical conductivity of GNG decreased with increasing N-doped content by carefully considering the electrical conductivity of N-doped carbon catalysts. Furthermore, PNG possessed a lower bandgap than GNG when the N content was higher than 2.8%, indicating that PNG had higher electrical conductivity (Fig. 2.17e). Therefore, it is concluded that electrical conductivity and Gibbs free energy of O_2 protonation are two important factors affecting ORR activity. In N-doped carbon materials with relatively high N content (> 2.8 at. %), the dominant active sites were enhanced by pyridine N due to their higher electrical conductivity. While for nitrogen-doped carbon materials with low nitrogen content (< 2.8 at. %), the conductivity difference between GNG and PNG was almost negligible. The initial ORR activity became the dominant factor for ORR. Therefore, in this case, graphitic N was considered to be the most effective moiety for ORR in N-doped carbon catalysts. This work discussed the possible reasons for the observation of so many different experimental results, suggesting new directions for studying active sites.

Besides N-doped carbon materials, other heteroatom (such as B, S, P, and F) doped carbon materials are also widely used as ORR catalysts. The B atoms with positive charge distribution are considered to be ORR active sites [208]. For F-doped carbon, F atoms as electron acceptors will not only improve the conductivity of F-doped carbon but also change its electronic structure [209]. In contrast, S-doped carbon interprets its ORR activity quite differently. S atoms have similar electronegativity to C atoms, so the change in atomic charge distribution is relatively small and electron spin is considered to be the most important factor contributing to the catalytic activity of ORR [210]. Specifically, when S atoms are incorporated into the carbon matrix,

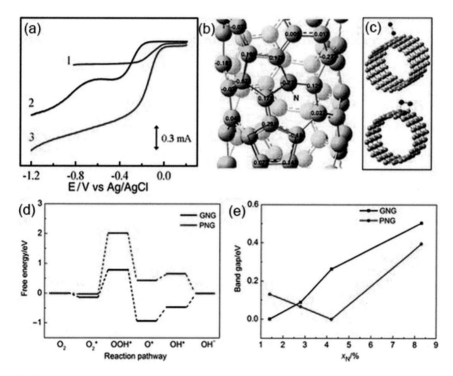

Fig. 2.17 a Pt/C/GC (curve 1), nitrogen-free and non-aligned carbon nanotubes (VA-CCNT/GC) supported by glassy carbon electrodes (curve 2) and aligned nitrogen-containing carbon nanotubes (VA-NCNT) (curve 3) Oxygen reduction RRDE voltammogram in air-saturated 0.1 M KOH. Due to technical difficulties related to sample mounting, the amperometric response of platinum ring electrodes with vertically aligned carbon nanotubes was not measured [11]. **b** Calculated charge density distribution of NCNTs. **c** Schematic illustration of possible adsorption modes of oxygen molecules on CCNTs (top) and NCNTs (bottom) [199]. **d** ORR free energy curves on GNG and PNG. **e** Bandgap and N-doping content of GNG and PNG [207]

the spins of adjacent carbon atoms will change from the previous symmetric type to the asymmetric type, resulting in stronger interaction between O_2 and carbon atoms. It is found that atoms with spin densities greater than 0.15 can effectively adsorb OOH or O_2 and become potential active sites for ORR. In addition, S atoms should be able to replace carbon atoms at the edge of graphene in the form of sulfur/sulfur oxide or connect two graphene sheets by forming sulfur cluster rings due to their smaller atomic size, thereby generating additional carbon active sites for ORR (Fig. 2.18a). Recently, based on systematic theoretical calculations, it was found that it was difficult for P atoms to directly replace C atoms in the sp^2 carbon lattice to form structures with three or four PC bonds (PC3G or PC4G). However, the P-doped graphene structure with PO bonds (OPC3G) was energetically more favorable and stable (Fig. 2.18b) [211]. The formation energy of OPC3G was rather negative, which was consistent with most XPS results and well confirms the existence of CP and CO. More importantly, it was noticed that OPC3G was most effective for ORR, where carbon atoms possessed a high negative charge act as active sites. This is the first discovery that carbon atoms with high negative charges can become active sites for ORR, which broadens the understanding of active sites on metal-free heteroatom-doped carbon.

2.4.3 Transition Metal-Modified Carbon-Based Catalysts

The incorporation of transition metals into carbon materials is another widely used strategy to achieve efficient ORR catalysts. In particular, doped carbon in the form of transition metals and N-C has received extensive attention. N-incorporated sp^2 carbons can act as electron donors or acceptors for transition metals. Furthermore, the highly porous structural carbon with a large surface area can enhance the exposure and mass transfer of active sites. For example, mesoporous hollow-spindle-shaped Fe-N-C catalysts were constructed by a novel "reactive hard-template" approach [213]. Fe_3C nanoparticles produced by polymerization on Fe_2O_3 nanospindle-shaped templates and subsequent calcination were encapsulated in N-doped graphitic layers. The easily accessible Fe-N sites and high mesoporous structure are the reasons for the high ORR activity. A straightforward and scalable molecularly confinement pyrolysis strategy was further developed to construct highly stable single Fe sites (3.5 wt% loadings) on N-doped carbon nanosheet catalysts [214]. Compared with commercial Pt/C, the synthesized catalyst exhibited superior ORR activity and stability in both alkaline and acidic environments. Furthermore, metal–organic frameworks (MOFs) can be used as self-templates for pyrolysis into transition metal-doped carbon materials with controllable porous structural features and variable elemental compositions [215–220]. For example, through multi-level structural optimization, a series of Co-Co_3O_4-based nanoarchitectures were constructed and embedded in hollow N-doped carbon polyhedra (Fig. 2.19) [221]. The optimal yolk@shell catalyst with abundant oxygen vacancies and efficient mass conversion and potential synergy of different

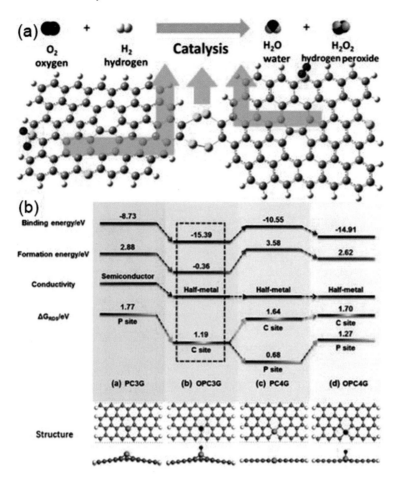

Fig. 2.18 **a** Possible active sites on S-doped carbon [212]. **b** Changes in binding energy, formation energy, conductivity, and ΔGRDS on four P-doped graphene models and their corresponding structures [208]

components presented competitive ORR activity and stability. However, these catalysts exhibited limited ORR performance in acidic environments, and the true active sites (M-N or M-C) remained unidentified. Therefore, an in-depth understanding of its active site and reaction mechanism can help to improve ORR efficiency.

In addition, Bao et al. found that metallic Fe nanoparticles encapsulated in peapod-type CNT interlayers exhibited stronger ORR activity compared with CNTs without Fe encapsulation through comprehensive experimental and DFT computational studies (Fig. 2.20a) [222, 223]. In addition, similar phenomena were observed in Co@NC [224–227], Fe_3C@NC [228–230], Fe_3C@C catalyst [175], etc. (Fig. 2.20b, c). It is believed that the work function of the carbon shell can be altered by the metal-core, resulting in favorable oxygen adsorption. Furthermore, this work function is

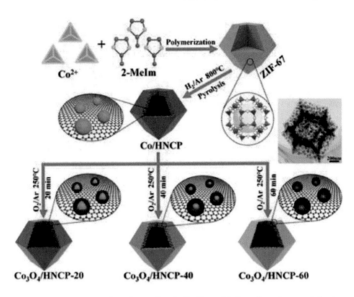

Fig. 2.19 Fabrication process of various Co_3O_4/hollow N-doped carbon polyhedra, and TEM images of the best catalyst (inset) [221]

thought to affect not only the carbon shell but also the metal-Nx active sites on the surface [231–235]. By comparing the catalysts with Fe/Fe_3C core or Fe-N_x structure, we found that Fe/Fe_3C nanocrystals improved the FeN$_x$ activity well (Fig. 2.20d) [236]. Further density functional theory (DFT) calculations revealed that the reduction of the Mulliken charge on the central iron atom of the FeN$_4$ structure could be regulated by additional iron atoms. The improved FeN$_4$ active sites have higher O_2 binding energy, which is beneficial to the adsorption of O_2, thereby promoting the ORR activity. Co@CoNC catalysts have been synthesized and a similar explanation has been proposed that the Co core can tune the electronic structure of the outermost nitrogen-doped carbon layer as well as the surface Co-N_x active sites [237]. Therefore, it is very important to develop new materials and techniques to study this structure, since catalysts obtained by high-temperature treatment are usually composed of such complex structures.

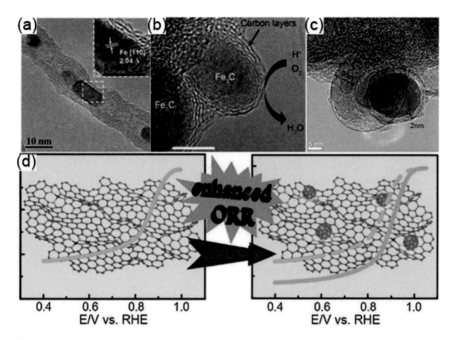

Fig. 2.20 a HRTEM image of Pod-Fe, inset showing the plane of Fe particles [223]. **b** Oxygen reduction process on Fe$_3$C@C (scale bar = 5.00 nm) [175]. **c** HRTEM image of Co@Co-N-C (scale bar = 5.00 nm). **d** The highly active Fe-N-C ORR catalyst model containing Fe-N$_x$ coordination sites and Fe/Fe$_3$C nanocrystals (Fe@C-FeNC) exhibits superior ORR activity due to the interaction between the encapsulated Fe/Fe$_3$C nanocrystals and Fe-N$_x$ [236]

References

1. P. Strasser, Acc. Chem. Res. **49**, 2658 (2016)
2. G. Wu, A. Santandreu, W. Kellogg, S. Gupta, O. Ogoke, H. Zhang, H.L. Wang, L. Dai, Nano Energy **29**, 83 (2016)
3. Z. Chen, J.Y. Choi, H. Wang, H. Li, Z. Chen, J. Power Sources **196**, 3673 (2011)
4. A. Li Zhu, H. Wang, W. Qu, X. Li, Z. Jong, H. Li, J. Power Sources, **195**, 5587 (2010)
5. G. Nam, J. Park, M. Choi, P. Oh, S. Park, M.G. Kim, N. Park, J. Cho, J.-S. Lee, ACS Nano **9**, 6493 (2015)
6. J.M. Ang, Y. Du, B.Y. Tay, C. Zhao, J. Kong, L.P. Stubbs, X. Lu, Langmuir **32**, 9265 (2016)
7. Q. Wu, L. Yang, X. Wang, Z. Hu, Acc. Chem. Res. **50**, 435 (2017)
8. W. Zhang, W. Lai, R. Cao, Chem. Rev. **117**, 3717 (2017)
9. Z. Xia, L. An, P. Chen, D. Xia, Adv. Energy Mater. **6**, 1600458 (2016)
10. G. Wu, K.L. More, C.M. Johnston, P. Zelenay, Science **332**, 443 (2011)
11. Y. Bing, H. Liu, L. Zhang, D. Ghosh, J. Zhang, Chem. Soc. Rev. **39**, 2184 (2010)
12. V.R. Stamenkovic, B. Fowler, B.S. Mun, G. Wang, P.N. Ross, C.A. Lucas, N.M. Markovic, Science **315**, 493 (2007)
13. X. Wang, Z. Li, Y. Qu, T. Yuan, W. Wang, Y. Wu, Y. Li, Chem **5**, 1486 (2019)
14. B.Y. Xia, W.T. Ng, H.B. Wu, X. Wang, X.W. Lou, Angew. Chem. Int. Ed. **51**, 7213 (2012)
15. X. Wang, Z. Chen, S. Chen, H. Wang, M. Huang, Chem. Eur. J. **26**, 12589 (2020)
16. P. Strasser, S. Koh, T. Anniyev, J. Greeley, K. More, C. Yu, Z. Liu, S. Kaya, D. Nordlund, H. Ogasawara, M.F. Toney, A. Nilsson, Nat. Chem. **2**, 454 (2010)

17. L. Wang, Z. Zeng, W. Gao, T. Maxson, D. Raciti, M. Giroux, X. Pan, C. Wang, J. Greeley, Science **363**, 870 (2019)
18. M. Li, Z. Zhao, T. Cheng, A. Fortunelli, C.Y. Chen, R. Yu, Q. Zhang, L. Gu, B.V. Merinov, Z. Lin, E. Zhu, T. Yu, Q. Jia, J. Guo, L. Zhang, W.A. Goddard, Y. Huang, X. Duan, Science **354**, 1414 (2016)
19. M. Du, L. Cui, Y. Cao, A.J. Bard, J. Am. Chem. Soc. **137**, 7397 (2015)
20. J. Zhu, A.O. Elnabawy, Z. Lyu, M. Xie, E.A. Murray, Z. Chen, W. Jin, M. Mavrikakis, Y. Xia, Mater. Today **35**, 69 (2020)
21. L. Chong, J. Wen, J. Kubal, F.G. Sen, J. Zou, J. Greeley, M. Chan, H. Barkholtz, W. Ding, D.J. Liu, Science **362**, 1276 (2018)
22. R. Tuinier, T.C. Lubensky, M. Escudero-Escribano, P. Malacrida, H.M. Hansen, U. Vej-Hansen, A. Velazquez-Palenzuela, V. Tripkovic, J. Schiøtz, J. Rossmeisl, I.E.L. Stephens, I. Chorkendorff, Science **352**, 73 (2016)
23. S. Li, J. Liu, Z. Yin, P. Ren, L. Lin, Y. Gong, C. Yang, X. Zheng, R. Cao, S. Yao, Y. Deng, X. Liu, L. Gu, W. Zhou, J. Zhu, X. Wen, B. Xu, D. Ma, ACS Catal. **10**, 907 (2020)
24. H. Huang, R. Chen, M. Liu, J. Wang, M.J. Kim, Z. Ye, Y. Xia, Top. Catal. **63**, 664 (2020)
25. X. Tian, X.F. Lu, B.Y. Xia, X.W. Lou, Joule **4**, 45 (2020)
26. X. Wang, S.I. Choi, L.T. Roling, M. Luo, C. Ma, L. Zhang, M. Chi, J. Liu, Z. Xie, J.A. Herron, M. Mavrikakis, Y. Xia, Nat. Commun. **6**, 7594 (2015)
27. X. Wang, L. Figueroa-Cosme, X. Yang, M. Luo, J. Liu, Z. Xie, Y. Xia, Nano Lett. **16**, 1467 (2016)
28. C.T. Lee, H. Wang, M. Zhao, T.H. Yang, M. Vara, Y. Xia, Chem. Eur. J. **25**, 5322 (2019)
29. M. Zhou, H. Wang, A.O. Elnabawy, Z.D. Hood, M. Chi, P. Xiao, Y. Zhang, M. Mavrikakis, Y. Xia, Chem. Mater. **31**, 1370 (2019)
30. K. Sasaki, H. Naohara, Y. Choi, Y. Cai, W.F. Chen, P. Liu, R.R. Adzic, Nat. Commun. **3**, 1115 (2012)
31. J. Li, H.M. Yin, X.B. Li, E. Okunishi, Y.L. Shen, J. He, Z.K. Tang, W.X. Wang, E. Yücelen, C. Li, Y. Gong, L. Gu, S. Miao, L.M. Liu, J. Luo, Y. Ding, Nat. Energy **2**, 17111 (2017)
32. A.L. Strickler, A. Jackson, T.F. Jaramillo, ACS Energy Lett. **2**, 244 (2017)
33. J. Zhu, M. Xie, Z. Chen, Z. Lyu, M. Chi, W. Jin, Y. Xia, Adv. Energy Mater. **10**, 1904114 (2020)
34. I.E.L. Stephens, A.S. Bondarenko, U. Grønbjerg, J. Rossmeisl, I. Chorkendorff, Energy Environ. Sci. **5**, 6744 (2012)
35. Y. Feng, Q. Shao, Y. Ji, X. Cui, Y. Li, X. Zhu, X. Huang, Sci. Adv. **4**, 8817 (2018)
36. V.R. Stamenkovic, B.S. Mun, M. Arenz, K.J.J. Mayrhofer, C.A. Lucas, G. Wang, P.N. Ross, N.M. Markovic, Nat. Mater. **6**, 241 (2007)
37. X. Shen, S. Dai, Y. Pan, L. Yao, J. Yang, X. Pan, J. Zeng, Z. Peng, ACS Catal. **9**, 11431 (2019)
38. M. Luo, S. Guo, Nat. Rev. Mater. **2**, 17059 (2017)
39. Y.J. Wang, N. Zhao, B. Fang, H. Li, X.T. Bi, H. Wang, Chem. Rev. **115**, 3433 (2015)
40. J. Pan, X.L. Tian, S. Zaman, Z. Dong, H. Liu, H.S. Park, B.Y. Xia, Batteries Supercaps **2**, 336 (2019)
41. F. Lin, K. Wang, Y. Tang, J. Lai, M. Lou, M. Huang, S. Guo, Chem. Commun. **54**, 1315 (2018)
42. F. Lu, J. Wang, J. Li, Y. Du, X.P. Kong, S. Liu, D. Yi, Y.K. Takahashi, K. Hono, X. Wang, J. Yao, Appl. Catal., B **278**, 119332 (2020)
43. M. Carlo, Science **343**, 1339 (2014)
44. B.N. Wanjala, B. Fang, J. Luo, Y. Chen, J. Yin, M.H. Engelhard, R. Loukrakpam, C.J. Zhong, J. Am. Chem. Soc. **133**, 12714 (2011)
45. Z. Meng, F. Xiao, Z. Wei, X. Guo, Y. Zhu, Y. Liu, G. Li, Z.Q. Yu, M. Shao, W.Y. Wong, Nano Res. **12**, 2954 (2019)
46. T.Y. Yoo, J.M. Yoo, A.K. Sinha, M.S. Bootharaju, E. Jung, H.S. Lee, B.H. Lee, J. Kim, W.H. Antink, Y.M. Kim, J. Lee, E. Lee, D.W. Lee, S.P. Cho, S.J. Yoo, Y.E. Sung, T. Hyeon, J. Am. Chem. Soc. **142**, 14190 (2020)
47. C. Cui, L. Gan, M. Heggen, S. Rudi, P. Strasser, Nat. Mater. **12**, 765 (2013)
48. R. Lin, X. Cai, H. Zeng, Z. Yu, Adv. Mater. **30**, 1705332 (2018)

49. J. Greeley, I.E.L. Stephens, A.S. Bondarenko, T.P. Johansson, H.A. Hansen, T.F. Jaramillo, J. Rossmeisl, I. Chorkendorff, J.K. Nørskov, Nat. Chem. **1**, 552 (2009)
50. P. Hernandez-Fernandez, F. Masini, D.N. McCarthy, C.E. Strebel, D. Friebel, D. Deiana, P. Malacrida, A. Nierhoff, A. Bodin, A.M. Wise, J.H. Nielsen, T.W. Hansen, A. Nilsson, I.E.L. Stephens, I. Chorkendorff, Nat. Chem. **6**, 732 (2014)
51. L. Zhang, K. Doyle-Davis, X. Sun, Energy Environ. Sci. **12**, 492 (2019)
52. Y. Chen, S. Ji, C. Chen, Q. Peng, D. Wang, Y. Li, Joule **2**, 1242 (2018)
53. J. Liu, M. Jiao, L. Lu, H.M. Barkholtz, Y. Li, L. Jiang, Z. Wu, D.J. Liu, L. Zhuang, C. Ma, J. Zeng, B. Zhang, D. Su, P. Song, W. Xing, W. Xu, Y. Wang, Z. Jiang, G. Sun, Nat. Commun. **8**, 15938 (2017)
54. T. Li, J. Liu, Y. Song, F. Wang, ACS Catal. **8**, 8450 (2018)
55. J. Kim, H.-E. Kim, H. Lee, Chem Sus Chem **11**, 104 (2018)
56. H. Zhang, G. Liu, L. Shi, J. Ye, Adv. Energy Mater. **8**, 1701343 (2018)
57. J. Liu, M. Jiao, B. Mei, Y. Tong, Y. Li, M. Ruan, P. Song, G. Sun, L. Jiang, Y. Wang, Z. Jiang, L. Gu, Z. Zhou, W. Xu, Angew. Chem. Int. Ed. **58**, 1163 (2019)
58. F. Calle-Vallejo, J.I. Martínez, J. Rossmeisl, Phys. Chem. Chem. Phys. **13**, 15639 (2011)
59. J.A.R. van Veen, J.F. van Baar, C.J. Kroese, J.G.F. Coolegem, N. de Wit, H.A. Colijn, Berichte der Bunsengesellschaft/Phys. Chem. Chem. Phys. **85**, 693 (1981)
60. M. Xiao, J. Zhu, G. Li, N. Li, S. Li, Z.P. Cano, L. Ma, P. Cui, P. Xu, G. Jiang, H. Jin, S. Wang, T. Wu, J. Lu, A. Yu, D. Su, Z. Chen, Angew. Chem. Int. Ed. **58**, 9640 (2019)
61. F. He, K. Li, C. Yin, Y. Wang, H. Tang, Z. Wu, Carbon **114**, 619 (2017)
62. Y. Xue, S. Sun, Q. Wang, Z. Dong, Z. Liu, J. Mater. Chem. A **6**, 10595 (2018)
63. J. Suntivich, H.A. Gasteiger, N. Yabuuchi, H. Nakanishi, J.B. Goodenough, Y. Shao-Horn, Nat. Chem. **3**, 546 (2011)
64. K. Chakrapani, G. Bendt, H. Hajiyani, T. Lunkenbein, M.T. Greiner, L. Masliuk, S. Salamon, J. Landers, R. Schlögl, H. Wende, R. Pentcheva, S. Schulz, M. Behrens, ACS Catal. **8**, 1259 (2018)
65. C. Mu, J. Mao, J. Guo, Q. Guo, Z. Li, W. Qin, Z. Hu, K. Davey, T. Ling, S.Z. Qiao, Adv. Mater. **32**, 1907168 (2020)
66. Y. Liang, Y. Li, H. Wang, J. Zhou, J. Wang, T. Regier, H. Dai, Nat. Mater. **10**, 780 (2011)
67. T. Li, B. Xue, B. Wang, G. Guo, D. Han, Y. Yan, A. Dong, J. Am. Chem. Soc. **139**, 12133 (2017)
68. J. Guo, Z. Mao, X. Yan, R. Su, P. Guan, B. Xu, X. Zhang, G. Qin, S.J. Pennycook, Nano Energy **28**, 261 (2016)
69. B. Anasori, M.R. Lukatskaya, Y. Gogotsi, Nat. Rev. Mater. **2**, 16098 (2017)
70. C. Ling, L. Shi, Y. Ouyang, Q. Chen, J. Wang, Adv. Sci. **3**, 1600180 (2016)
71. Y. Yu, J. Zhou, Z. Sun, Adv. Funct. Mater. **30**, 2000570 (2020)
72. X. Ma, H. Meng, M. Cai, P.K. Shen, J. Am. Chem. Soc. **2012**, 134 (1954)
73. J. Gautam, T.D. Thanh, K. Maiti, N.H. Kim, J.H. Lee, Carbon **137**, 358 (2018)
74. P. Żółtowski, D.M. Dražić, L. Vorkapić, Journal **3**, 271 (1973)
75. Y. Meng, W. Song, H. Huang, Z. Ren, S.Y. Chen, S.L. Suib, J. Am. Chem. Soc. **136**, 11452 (2014)
76. F. Cheng, Y. Su, J. Liang, Z. Tao, J. Chen, Chem. Mater. **22**, 898 (2010)
77. L. Mao, D. Zhang, T. Sotomura, K. Nakatsu, N. Koshiba, T. Ohsaka, Electrochim. Acta **48**, 1015 (2003)
78. K.L. Pickrahn, S.W. Park, Y. Gorlin, H.B.R. Lee, T.F. Jaramillo, S.F. Bent, Adv. Energy Mater. **2**, 1269 (2012)
79. Y. Gorlin, C.J. Chung, D. Nordlund, B.M. Clemens, T.F. Jaramillo, ACS Catal. **2**, 2687 (2012)
80. F. Cheng, J. Shen, W. Ji, Z. Tao, J. Chen, ACS Appl. Mater. Interfaces **1**, 460 (2009)
81. F.H.B. Lima, M.L. Calegaro, E.A. Ticianelli, J. Electroanal. Chem. **590**, 152 (2006)
82. S. Lee, G. Nam, J. Sun, J.S. Lee, H.W. Lee, W. Chen, J. Cho, Y. Cui, Angew. Chem. Int. Ed. **55**, 8599 (2016)
83. S. Zhao, T. Liu, D. Shi, Y. Zhang, W. Zeng, T. Li, B. Miao, Appl. Surf. Sci. **351**, 862 (2015)

84. G. Cheng, S. Xie, B. Lan, X. Zheng, F. Ye, M. Sun, X. Lu, L. Yu, J. Mater. Chem. A **4**, 16462 (2016)
85. S. Ghosh, P. Kar, N. Bhandary, S. Basu, S. Sardar, T. Maiyalagan, D. Majumdar, S.K. Bhattacharya, A. Bhaumik, P. Lemmens, S.K. Pal, Catal. Sci. Technol. **6**, 1417 (2016)
86. M.C. Wu, T.S. Zhao, H.R. Jiang, L. Wei, Z.H. Zhang, Electrochim. Acta **222**, 1438 (2016)
87. K. Lei, L. Cong, X. Fu, F. Cheng, J. Chen, Inorg. Chem. Front. **3**, 928 (2016)
88. M. Lehtimäki, H. Hoffmannová, O. Boytsova, Z. Bastl, M. Busch, N.B. Halck, J. Rossmeisl, P. Krtil, Electrochim. Acta **191**, 452 (2016)
89. T.T. Truong, Y. Liu, Y. Ren, L. Trahey, Y. Sun, ACS Nano **6**, 8067 (2012)
90. K. Selvakumar, S.M. Senthil Kumar, R. Thangamuthu, K. Ganesan, P. Murugan, P. Rajput, S.N. Jha, D. Bhattacharyya, J. Phys. Chem. C **119**, 6604 (2015)
91. J. Liu, L. Jiang, T. Zhang, J. Jin, L. Yuan, G. Sun, Electrochim. Acta **205**, 38 (2016)
92. C. Jiang, Z. Guo, Y. Zhu, H. Liu, M. Wan, L. Jiang, Chemsuschem **8**, 158 (2015)
93. C. Wei, L. Yu, C. Cui, J. Lin, C. Wei, N. Mathews, F. Huo, T. Sritharan, Z. Xu, Chem. Commun. **50**, 7885 (2014)
94. W. Wang, J. Geng, L. Kuai, M. Li, B. Geng, Chem. Eur. J. **22**, 9909 (2016)
95. T. Zhang, F. Cheng, J. Du, Y. Hu, J. Chen, Adv. Energy Mater. **5**, 1400654 (2015)
96. S. Sun, H. Miao, Y. Xue, Q. Wang, S. Li, Z. Liu, Electrochim. Acta **214**, 49 (2016)
97. Q. Tang, L. Jiang, J. Qi, Q. Jiang, S. Wang, G. Sun, Appl. Catal., B, **104**, 337 (2011)
98. J. Liu, J. Liu, W. Song, F. Wang, Y. Song, J. Mater. Chem. A **2**, 17477 (2014)
99. S.A. Park, H. Lim, Y.T. Kim, ACS Catal. **5**, 3995 (2015)
100. G. Chen, J. Sunarso, Y. Zhu, J. Yu, Y. Zhong, W. Zhou, Z. Shao, ChemElectroChem **3**, 1760 (2016)
101. H. Cheng, K. Xu, L. Xing, S. Liu, Y. Gong, L. Gu, L. Zhang, C. Wu, J. Mater. Chem. A **4**, 11775 (2016)
102. H. Chai, J. Xu, J. Han, Y. Su, Z. Sun, D. Jia, W. Zhou, J. Colloid Interface Sci. **488**, 251 (2017)
103. Y.G. Wang, L. Cheng, F. Li, H.M. Xiong, Y.Y. Xia, Chem. Mater. **19**, 2095 (2007)
104. J. Duan, Y. Zheng, S. Chen, Y. Tang, M. Jaroniec, S. Qiao, Chem. Commun. **49**, 7705 (2013)
105. S.K. Bikkarolla, F. Yu, W. Zhou, P. Joseph, P. Cumpson, P. Papakonstantinou, J. Mater. Chem. A **2**, 14493 (2014)
106. K.H. Wu, Q. Zeng, B. Zhang, X. Leng, D.S. Su, I.R. Gentle, D.W. Wang, ChemSusChem **8**, 3331 (2015)
107. Y. Wang, X. Ding, F. Wang, J. Li, S. Song, H. Zhang, Chem. Sci. **7**, 4284 (2016)
108. Q. Xue, Z. Pei, Y. Huang, M. Zhu, Z. Tang, H. Li, Y. Huang, N. Li, H. Zhang, C. Zhi, J. Mater. Chem. A **5**, 20818 (2017)
109. Y.Q. Lyu, C. Chen, Y. Gao, M. Saccoccio, F. Ciucci, J. Mater. Chem. A **4**, 19147 (2016)
110. S. Sun, H. Miao, Y. Xue, Q. Wang, Q. Zhang, Z. Dong, S. Li, H. Huang, Z. Liu, J. Electrochem. Soc. **164**, F768 (2017)
111. T.N. Lambert, J.A. Vigil, S.E. White, C.J. Delker, D.J. Davis, M. Kelly, M.T. Brumbach, M.A. Rodriguez, B.S. Swartzentruber, J. Phys. Chem. C **121**, 2789 (2017)
112. Y. Gorlin, T.F. Jaramillo, J. Am. Chem. Soc. **132**, 13612 (2010)
113. Q. Wu, L. Jiang, L. Qi, E. Wang, G. Sun, Int. J. Hydrogen Energy **39**, 3423 (2014)
114. J. Zhang, Z. Lyu, F. Zhang, L. Wang, P. Xiao, K. Yuan, M. Lai, W. Chen, J. Mater. Chem. A **4**, 6350 (2016)
115. J. Zhou, T. Ge, X. Cui, J. Lv, H. Guo, Z. Hua, J. Shi, ChemElectroChem **4**, 1279 (2017)
116. T. Odedairo, X. Yan, J. Ma, Y. Jiao, X. Yao, A. Du, Z. Zhu, ACS Appl. Mater. Interfaces **7**, 21373 (2015)
117. J. Wang, R. Gao, D. Zhou, Z. Chen, Z. Wu, G. Schumacher, Z. Hu, X. Liu, ACS Catal. **7**, 6533 (2017)
118. L. Liu, H. Guo, Y. Hou, J. Wang, L. Fu, J. Chen, H. Liu, J. Wang, Y. Wu, J. Mater. Chem. A **5**, 14673 (2017)
119. L. Leng, X. Zeng, H. Song, T. Shu, H. Wang, S. Liao, J. Mater. Chem. A **3**, 15626 (2015)
120. L. Liu, Y. Hou, J. Wang, J. Chen, H.K. Liu, Y. Wu, J. Wang, Adv. Mater. Interfaces **3**, 1600030 (2016)

121. Y. He, J. Zhang, G. He, X. Han, X. Zheng, C. Zhong, W. Hu, Y. Deng, Nanoscale **9**, 8623 (2017)
122. Q. Zhao, Z. Yan, C. Chen, J. Chen, Chem. Rev. **117**, 10121 (2017)
123. C. Guan, A. Sumboja, H. Wu, W. Ren, X. Liu, H. Zhang, Z. Liu, C. Cheng, S.J. Pennycook, J. Wang, Adv. Mater. **29**, 1704117 (2017)
124. M. Yu, Z. Wang, C. Hou, Z. Wang, C. Liang, C. Zhao, Y. Tong, X. Lu, S. Yang, Adv. Mater. **29**, 1602868 (2017)
125. J. Li, Z. Zhou, K. Liu, F. Li, Z. Peng, Y. Tang, H. Wang, J. Power Sources **343**, 30 (2017)
126. A. Aijaz, J. Masa, C. Rösler, W. Xia, P. Weide, A.J.R. Botz, R.A. Fischer, W. Schuhmann, M. Muhler, Angew. Chem. Int. Ed. **55**, 4087 (2016)
127. J.E. Houston, G.E. Laramore, R.L. Park, Science **185**, 258 (1974)
128. L. Dai, M. Liu, Y. Song, J. Liu, F. Wang, Nano Energy **27**, 185 (2016)
129. A.A. Ensafi, Z. Ahmadi, M. Jafari-Asl, B. Rezaei, Electrochim. Acta **173**, 619 (2015)
130. Z. Cui, S. Wang, Y. Zhang, M. Cao, J. Power Sources **272**, 808 (2014)
131. X. Zhang, Q. Xiao, Y. Zhang, X. Jiang, Z. Yang, Y. Xue, Y.M. Yan, K. Sun, J. Phys. Chem. C **118**, 20229 (2014)
132. N. Xu, J. Qiao, X. Zhang, C. Ma, S. Jian, Y. Liu, P. Pei, Appl. Energy **175**, 495 (2016)
133. W. Gu, J. Liu, M. Hu, F. Wang, Y. Song, ACS Appl. Mater. Interfaces **7**, 26914 (2015)
134. Y. Su, H. Jiang, Y. Zhu, X. Yang, J. Shen, W. Zou, J. Chen, C. Li, J. Mater. Chem. A **2**, 7281 (2014)
135. Z.-S. Wu, S. Yang, Y. Sun, K. Parvez, X. Feng, K. Müllen, J. Am. Chem. Soc. **134**, 9082 (2012)
136. V.K. Singh, M.K. Patra, M. Manoth, G.S. Gowd, S.R. Vadera, N. Kumar, Xinxing Tan Cailiao/New Carbon Mater. **24**, 147 (2009)
137. Z. Ma, X. Huang, S. Dou, J. Wu, S. Wang, J. Phys. Chem. C **118**, 17231 (2014)
138. P.A. Pascone, D. Berk, J.L. Meunier, Catal. Today **211**, 162 (2013)
139. S. Parwaiz, K. Bhunia, A.K. Das, M.M. Khan, D. Pradhan, J. Phys. Chem. C **121**, 20165 (2017)
140. S. Sun, Y. Xue, Q. Wang, S. Li, H. Huang, H. Miao, Z. Liu, Chem. Commun. **53**, 7921 (2017)
141. V. Vij, S. Sultan, A.M. Harzandi, A. Meena, J.N. Tiwari, W.G. Lee, T. Yoon, K.S. Kim, ACS Catal. **7**, 7196 (2017)
142. W. Xia, A. Mahmood, Z. Liang, R. Zou, S. Guo, Angew. Chem. Int. Ed. **55**, 2650 (2016)
143. M. Sun, H. Liu, Y. Liu, J. Qu, J. Li, Nanoscale **7**, 1250 (2015)
144. D.B. Meadowcroft, Nature **226**, 847 (1970)
145. Y. Matumoto, H. Yoneyama, H. Tamura, Chem. Lett. **4**, 661 (1975)
146. Y. Matsumoto, H. Yoneyama, H. Tamura, J. Electroanal. Chem. **79**, 319 (1977)
147. Y. Yuan, J. Wang, S. Adimi, H. Shen, T. Thomas, R. Ma, J.P. Attfield, M. Yang, Nat. Mater. **19**, 282 (2020)
148. Y. Dong, Y. Deng, J. Zeng, H. Song, S. Liao, J. Mater. Chem. A **5**, 5829 (2017)
149. L. An, Y. Li, M. Luo, J. Yin, Y.Q. Zhao, C. Xu, F. Cheng, Y. Yang, P. Xi, S. Guo, Adv. Funct. Mater. **27**, 1703779 (2017)
150. H.-F. Wang, C. Tang, B. Wang, B.-Q. Li, Q. Zhang, Adv. Mater. **29**, 1702327 (2017)
151. L. Li, L. Song, H. Guo, W. Xia, C. Jiang, B. Gao, C. Wu, T. Wang, J. He, Nanoscale **11**, 901 (2019)
152. V.V.T. Doan-Nguyen, S. Zhang, E.B. Trigg, R. Agarwal, J. Li, D. Su, K.I. Winey, C.B. Murray, ACS Nano **9**, 8108 (2015)
153. K.P. Singh, E.J. Bae, J.S. Yu, J. Am. Chem. Soc. **137**, 3165 (2015)
154. A. Parra-Puerto, K.L. Ng, K. Fahy, A.E. Goode, M.P. Ryan, A. Kucernak, ACS Catal. **9**, 11515 (2019)
155. R. Zhang, C. Zhang, W. Chen, J. Mater. Chem. A **4**, 18723 (2016)
156. D.C. Nguyen, D.T. Tran, T.L. Luyen Doan, N.H. Kim, J.H. Lee, Chem. Mater., **31**, 2892 (2019)
157. S. Chakrabartty, D. Sahu, C.R. Raj, ACS Appl. Energy Mater. **3**, 2811 (2020)
158. M.M. Kumar, C.R. Raj, ACS Appl. Nano Mater. **2**, 643 (2019)

159. L. Chen, Y. Zhang, L. Dong, X. Liu, L. Long, S. Wang, C. Liu, S. Dong, J. Jia, Carbon **158**, 885 (2020)
160. R. Wang, Y. Yuan, J. Zhang, X. Zhong, J. Liu, Y. Xie, S. Zhong, Z. Xu, J. Power Sources **501**, 230006 (2021)
161. R.E. Witkowski, W.G. Fateley, 1291 (1966)
162. J. Xie, Y. Xie, Chem. Eur. J. **22**, 3588 (2016)
163. M.E. Kreider, M.B. Stevens, Y. Liu, A.M. Patel, M.J. Statt, B.M. Gibbons, A. Gallo, M. Ben-Naim, A. Mehta, R.C. Davis, A.V. Ievlev, J.K. Nørskov, R. Sinclair, L.A. King, T.F. Jaramillo, Chem. Mater. **32**, 2946 (2020)
164. V.G. Anju, R. Manjunatha, P.M. Austeria, S. Sampath, J. Mater. Chem. A **4**, 5258 (2016)
165. J. Luo, X. Qiao, J. Jin, X. Tian, H. Fan, D. Yu, W. Wang, S. Liao, N. Yu, Y. Deng, J. Mater. Chem. A **8**, 8575 (2020)
166. S. Mondal, C.R. Raj, J. Phys. Chem. C **122**, 18468 (2018)
167. Y. Yang, R. Zeng, Y. Xiong, F.J. Disalvo, H.D. Abruña, J. Am. Chem. Soc. **141**, 19241 (2019)
168. F. Liu, X. Yang, D. Dang, X. Tian, ChemElectroChem **6**, 2208 (2019)
169. H. Nan, D. Dang, X.L. Tian, J. Mater. Chem. A **6**, 6065 (2018)
170. X. Tian, J. Luo, H. Nan, Z. Fu, J. Zeng, S. Liao, J. Mater. Chem. A **3**, 16801 (2015)
171. X.L. Tian, L. Wang, B. Chi, Y. Xu, S. Zaman, K. Qi, H. Liu, S. Liao, B.Y. Xia, ACS Catal. **8**, 8970 (2018)
172. J.G. Chen, Chem. Rev. **96**, 1477 (1996)
173. T. Iida, M. Shetty, K. Murugappan, Z. Wang, K. Ohara, T. Wakihara, Y. Román-Leshkov, ACS Catal. **7**, 8147 (2017)
174. S.T. Hunt, T. Nimmanwudipong, Y. Román-Leshkov, Angew. Chem. Int. Ed. **53**, 5131 (2014)
175. Y. Hu, J.O. Jensen, W. Zhang, L.N. Cleemann, W. Xing, N.J. Bjerrum, Q. Li, Angew. Chem. Int. Ed. **53**, 3675 (2014)
176. M. Xiao, J. Zhu, L. Feng, C. Liu, W. Xing, Adv. Mater. **27**, 2521 (2015)
177. S. Dou, L. Tao, J. Huo, S. Wang, L. Dai, Energy Environ. Sci. **9**, 1320 (2016)
178. S. Yang, J. Kim, Y.J. Tak, A. Soon, H. Lee, Angew. Chem. Int. Ed. **55**, 2058 (2016)
179. Y. Chen, S. Ji, Y. Wang, J. Dong, W. Chen, Z. Li, R. Shen, L. Zheng, Z. Zhuang, D. Wang, Y. Li, Angew. Chem. Int. Ed. **56**, 6937 (2017)
180. H. Zhang, H.T. Chung, D.A. Cullen, S. Wagner, U.I. Kramm, K.L. More, P. Zelenay, G. Wu, Energy Environ. Sci. **12**, 2548 (2019)
181. P. Chen, T. Zhou, L. Xing, K. Xu, Y. Tong, H. Xie, L. Zhang, W. Yan, W. Chu, C. Wu, Y. Xie, Angew. Chem. Int. Ed. **56**, 610 (2017)
182. H. Zhang, S. Hwang, M. Wang, Z. Feng, S. Karakalos, L. Luo, Z. Qiao, X. Xie, C. Wang, D. Su, Y. Shao, G. Wu, J. Am. Chem. Soc. **139**, 14143 (2017)
183. J. Li, H. Zhang, W. Samarakoon, W. Shan, D.A. Cullen, S. Karakalos, M. Chen, D. Gu, K.L. More, G. Wang, Z. Feng, Z. Wang, G. Wu, Angew. Chem. Int. Ed. **58**, 18971 (2019)
184. C. Wang, J. Kim, J. Tang, M. Kim, H. Lim, V. Malgras, J. You, Q. Xu, J. Li, Y. Yamauchi, Chem **6**, 19 (2020)
185. X. Wan, X. Liu, Y. Li, R. Yu, L. Zheng, W. Yan, H. Wang, M. Xu, J. Shui, Nat. Catal. **2**, 259 (2019)
186. Q. Li, W. Chen, H. Xiao, Y. Gong, Z. Li, L. Zheng, X. Zheng, W. Yan, W.C. Cheong, R. Shen, N. Fu, L. Gu, Z. Zhuang, C. Chen, D. Wang, Q. Peng, J. Li, Y. Li, Adv. Mater. **30**, 1800588 (2018)
187. Y. He, S. Hwang, D.A. Cullen, M.A. Uddin, L. Langhorst, B. Li, S. Karakalos, A.J. Kropf, E.C. Wegener, J. Sokolowski, M. Chen, D. Myers, D. Su, K.L. More, G. Wang, S. Litster, G. Wu, Energy Environ. Sci. **12**, 250 (2019)
188. X. Xiong, X. Li, Y. Jia, Y. Meng, K. Sun, L. Zheng, G. Zhang, Y. Li, X. Sun, Nanoscale **11**, 15900 (2019)
189. L. Bai, Z. Duan, X. Wen, R. Si, J. Guan, Appl. Catal., B **257**, 117930 (2019)
190. Y. Yang, K. Mao, S. Gao, H. Huang, G. Xia, Z. Lin, P. Jiang, C. Wang, H. Wang, Q. Chen, Adv. Mater. **30**, 1801732 (2018)

191. J. Li, M. Chen, D.A. Cullen, S. Hwang, M. Wang, B. Li, K. Liu, S. Karakalos, M. Lucero, H. Zhang, C. Lei, H. Xu, G.E. Sterbinsky, Z. Feng, D. Su, K.L. More, G. Wang, Z. Wang, G. Wu, Nat. Catal. **1**, 935 (2018)
192. M. Seredych, A. Szczurek, V. Fierro, A. Celzard, T.J. Bandosz, ACS Catal. **6**, 5618 (2016)
193. M.R. Benzigar, S. Joseph, H. Ilbeygi, D.H. Park, S. Sarkar, G. Chandra, S. Umapathy, S. Srinivasan, S.N. Talapaneni, A. Vinu, Angew. Chem. Int. Ed. **57**, 569 (2018)
194. Y. Jiang, L. Yang, T. Sun, J. Zhao, Z. Lyu, O. Zhuo, X. Wang, Q. Wu, J. Ma, Z. Hu, ACS Catal. **5**, 6707 (2015)
195. L. Tao, Q. Wang, S. Dou, Z. Ma, J. Huo, S. Wang, L. Dai, Chem. Commun. **52**, 2764 (2016)
196. A. Shen, Y. Zou, Q. Wang, R.A.W. Dryfe, X. Huang, S. Dou, L. Dai, S. Wang, Angew. Chem. Int. Ed. **53**, 10804 (2014)
197. W. Ding, L. Li, K. Xiong, Y. Wang, W. Li, Y. Nie, S. Chen, X. Qi, Z. Wei, J. Am. Chem. Soc. **137**, 5414 (2015)
198. L. Yang, J. Shui, L. Du, Y. Shao, J. Liu, L. Dai, Z. Hu, Adv. Mater. **31**, 1804799 (2019)
199. K. Gong, F. Du, Z. Xia, M. Durstock, L. Dai, Science **323**, 760 (2009)
200. Y. Zhao, J. Wan, H. Yao, L. Zhang, K. Lin, L. Wang, N. Yang, D. Liu, L. Song, J. Zhu, L. Gu, L. Liu, H. Zhao, Y. Li, D. Wang, Nat. Chem. **10**, 924 (2018)
201. D. Yan, Y. Li, J. Huo, R. Chen, L. Dai, S. Wang, Adv. Mater. **29**, 1606459 (2017)
202. C. Hu, L. Dai, Adv. Mater. **29**, 1604942 (2017)
203. D. Guo, R. Shibuya, C. Akiba, S. Saji, T. Kondo, J. Nakamura, Science **351**, 361 (2016)
204. H.B. Yang, J. Miao, S.-F. Hung, J. Chen, H.B. Tao, X. Wang, L. Zhang, R. Chen, J. Gao, H.M. Chen, L. Dai, B. Liu, Sci. Adv. **2**, e1501122 (2016)
205. L. Lai, J.R. Potts, D. Zhan, L. Wang, C.K. Poh, C. Tang, H. Gong, Z. Shen, J. Lin, R.S. Ruoff, Energy Environ. Sci. **5**, 7936 (2012)
206. T. Sharifi, G. Hu, X. Jia, T. Wågberg, ACS Nano **6**, 8904 (2012)
207. J. Wang, L. Li, Z.-D. Wei, Acta Phys. Chim. Sin. **32**, 321 (2016)
208. L. Wang, H. Dong, Z. Guo, L. Zhang, T. Hou, Y. Li, J. Phys. Chem. C **120**, 17427 (2016)
209. G. Panomsuwan, N. Saito, T. Ishizaki, J. Mater. Chem. A **3**, 9972 (2015)
210. I.Y. Jeon, S. Zhang, L. Zhang, H.J. Choi, J.M. Seo, Z. Xia, L. Dai, J.B. Baek, Adv. Mater. **25**, 6138 (2013)
211. N. Yang, X. Zheng, L. Li, J. Li, Z. Wei, J. Phys. Chem. C **121**, 19321 (2017)
212. L. Zhang, J. Niu, M. Li, Z. Xia, J. Phys. Chem. C **118**, 3545 (2014)
213. X. Xin, H. Qin, H.P. Cong, S.H. Yu, Langmuir **34**, 4952 (2018)
214. Z. Yang, Y. Wang, M. Zhu, Z. Li, W. Chen, W. Wei, T. Yuan, Y. Qu, Q. Xu, C. Zhao, X. Wang, P. Li, Y. Li, Y. Wu, Y. Li, ACS Catal. **9**, 2158 (2019)
215. X.F. Lu, B.Y. Xia, S.Q. Zang, X.W. Lou, Angew. Chem. Int. Ed. **59**, 4634 (2020)
216. W. Li, S. Xue, S. Watzele, S. Hou, J. Fichtner, A.L. Semrau, L. Zhou, A. Welle, A.S. Bandarenka, R.A. Fischer, Angew. Chem. Int. Ed. **59**, 5837 (2020)
217. F. Yang, P.L. Deng, Q. Wang, J. Zhu, Y. Yan, L. Zhou, K. Qi, H. Liu, H.S. Park, B.Y. Xia, J. Mater. Chem. A **8**, 12418 (2020)
218. P. Deng, F. Yang, Z. Wang, S. Chen, Y. Zhou, S. Zaman, B.Y. Xia, Angew. Chem. Int. Ed. **59**, 10807 (2020)
219. Q. Wang, Y. Luo, R. Hou, S. Zaman, K. Qi, H. Liu, H.S. Park, B.Y. Xia, Adv. Mater. **31**, 1905744 (2019)
220. F. Yang, A. Chen, P.L. Deng, Y. Zhou, Z. Shahid, H. Liu, B.Y. Xia, Chem. Sci. **10**, 7975 (2019)
221. D. Ding, K. Shen, X. Chen, H. Chen, J. Chen, T. Fan, R. Wu, Y. Li, ACS Catal. **8**, 7879 (2018)
222. D. Deng, L. Yu, X. Chen, G. Wang, L. Jin, X. Pan, J. Deng, G. Sun, X. Bao, Angew. Chem. Int. Ed. **52**, 371 (2013)
223. J. Deng, L. Yu, D. Deng, X. Chen, F. Yang, X. Bao, J. Mater. Chem. A **1**, 14868 (2013)
224. L.B. Lv, T.N. Ye, L.H. Gong, K.X. Wang, J. Su, X.H. Li, J.S. Chen, Chem. Mater. **27**, 544 (2015)
225. M. Zeng, Y. Liu, F. Zhao, K. Nie, N. Han, X. Wang, W. Huang, X. Song, J. Zhong, Y. Li, Adv. Funct. Mater. **26**, 4397 (2016)
226. D. Li, C. Lv, L. Liu, Y. Xia, X. She, S. Guo, D. Yang, ACS Cent. Sci. **1**, 261 (2015)

227. J. Meng, C. Niu, L. Xu, J. Li, X. Liu, X. Wang, Y. Wu, X. Xu, W. Chen, Q. Li, Z. Zhu, D. Zhao, L. Mai, J. Am. Chem. Soc. **139**, 8212 (2017)
228. J.A. Varnell, E.C.M. Tse, C.E. Schulz, T.T. Fister, R.T. Haasch, J. Timoshenko, A.I. Frenkel, A.A. Gewirth, Nat. Commun. **7**, 12582 (2016)
229. Y. Zhang, W.J. Jiang, L. Guo, X. Zhang, J.S. Hu, Z. Wei, L.J. Wan, ACS Appl. Mater. Interfaces **7**, 11508 (2015)
230. Z. Wen, S. Ci, F. Zhang, X. Feng, S. Cui, S. Mao, S. Luo, Z. He, J. Chen, Adv. Mater. **24**, 1399 (2012)
231. H. Jiang, Y. Yao, Y. Zhu, Y. Liu, Y. Su, X. Yang, C. Li, ACS Appl. Mater. Interfaces **7**, 21511 (2015)
232. J.H. Kim, Y.J. Sa, H.Y. Jeong, S.H. Joo, ACS Appl. Mater. Interfaces **9**, 9567 (2017)
233. G. Ren, X. Lu, Y. Li, Y. Zhu, L. Dai, L. Jiang, ACS Appl. Mater. Interfaces **8**, 4118 (2016)
234. S. Zhou, N. Liu, Z. Wang, J. Zhao, ACS Appl. Mater. Interfaces **9**, 22578 (2017)
235. Z.Y. Wu, X.X. Xu, B.C. Hu, H.W. Liang, Y. Lin, L.F. Chen, S.H. Yu, Angew. Chem. Int. Ed. **54**, 8179 (2015)
236. W.J. Jiang, L. Gu, L. Li, Y. Zhang, X. Zhang, L.J. Zhang, J.Q. Wang, J.S. Hu, Z. Wei, L.J. Wan, J. Am. Chem. Soc. **138**, 3570 (2016)
237. Y. Wang, Y. Nie, W. Ding, S.G. Chen, K. Xiong, X.Q. Qi, Y. Zhang, J. Wang, Z.D. Wei, Chem. Commun. **51**, 8942 (2015)

Chapter 3
Cathode Materials for Secondary Zinc-Air Batteries

3.1 Overview

The electrochemical reduction of oxygen and the evolution of oxygen are two important electrode reactions that have been extensively studied to develop electrochemical energy conversion and storage techniques based on oxygen electrocatalysis (Fig. 3.1). The exploitation of inexpensive, highly active and durable non-precious metal-based oxygen electrocatalyst is essential for emerging energy technologies, including anion exchange membrane fuel cells, metal-air batteries (MABs), water electrolyzers, etc. [1].

The activity of the oxygen electrocatalysts largely determines the overall energy storage performance of these devices. Although platinum and ruthenium/iridium-based catalysts are known for their catalytic activity towards oxygen reduction reaction (ORR) and oxygen evolution reaction (OER), the high cost and lack of durability limit their widespread use in practical applications. This chapter discusses the research progress of ORR/OER bifunctional oxygen electrocatalytic activity, including noble metal-based, transition metal-based, and carbon-based catalysts, and their applications in zinc-air batteries (ZABs). These electrocatalysts require rational surface and chemical engineering to achieve the desired oxygen electrocatalytic activity. Surface engineering increases the number of active sites, while chemical engineering increases the intrinsic activity of the catalysts. The encapsulation or integration of the active catalysts with undoped or heteroatom-doped carbon nanostructures results in enhanced durability of the active catalysts. In many cases, the synergistic effect between the heteroatom-doped carbon matrix and the active catalyst plays an important role in controlling the catalytic activity. The ORR activities of these catalysts are evaluated in terms of onset potential, the number of electrons transferred, limiting current density, and durability. On the other hand, bifunctional oxygen electrocatalytic activity and ZAB performance are measured as the potential difference between ORR and OER, $\Delta E = E_{j10\ OER} - E_{1/2\ ORR}$, specific capacity, peak power density, open-circuit voltage, volt efficiency and charge/discharge cycle stability. ZABs based on non-noble metal electrocatalysts are promising, which can

S. Peng, *Zinc-Air Batteries*, https://doi.org/10.1007/978-981-19-8214-9_3

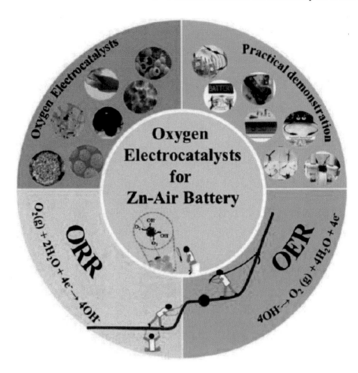

Fig. 3.1 Bifunctional oxygen catalyst and its significance for zinc-air batteries

provide high power density, specific capacity and round-trip efficiency. The active sites of oxygen electrocatalysis and challenges related to carbon support are briefly discussed. Although considerable progress has been made in the emerging electrocatalysts in recent years, several problems remain to be solved to realize the commercial potential of the practical application of rechargeable ZAB.

3.2   Precious Metal Bifunctional Catalysts

Precious metals, such as Pt and Ir, are well known as benchmark catalysts for ORR and OER, respectively. Although they have the advantages of high catalytic activity and electrical conductivity, their high cost and scarcity are serious problems, so it is necessary to optimize their activities by changing the morphology and composition of noble metal catalysts. Another feasible strategy is to alloy or modify the noble metals with other suitable transition metals to improve performance and reduce catalyst cost simultaneously.

3.2.1 Precious Metal Alloy Catalysts

Over the past decades, many strategies have been proposed to improve the performance of precious metal catalysts. For example, tuning size and morphology to achieve efficient catalytic surfaces with small dispersion size and high surface area has proven to be an effective way to improve ORR performance.

The precious metal platinum (Pt) is the most effective catalyst for ORR reactions due to its high stability and superior electrocatalytic activity. S. Ravichandran et al. synthesized Pt-Ru-Ir mesoporous nanostructures by a chemical reduction strategy with KIT-6 mesoporous silicon as a template (Fig. 3.2) [2]. The synergy of different elements in the alloy, the electronic effect, and the lattice stress induced during the alloy formation process promote its OER/ORR bifunctional catalytic ability. The ORR mass activity reached 0.21 mA μg^{-1} with an E_{10} of 470 mV.

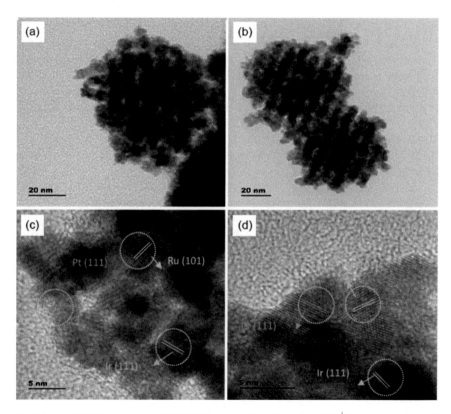

Fig. 3.2 **a, b** TEM images and **c, d** HRTEM images of $Pt_{70}Ru_{20}Ir_{10}$ and $Pt_{70}Ru_{15}Ir_{15}$ catalysts [2]

3.2.2 Precious Metal Composite Carbon Carrier Catalysts

Precious metal-based catalysts are usually well dispersed on high surface area carbon carriers (e.g., carbon black, carbon nanotubes (CNTs) and graphene). Silver-based catalysts are often combined with carbon carriers as ORR/OER active composites.

A nitrogen-doped noble metal biomass C-Ag composite was synthesized using soymilk as nitrogen and carbon precursor materials. After heat-treated at 900 °C, the ORR onset potential was 0.93 V versus RHE, equivalent to 20 wt% Pt/C. With only 10 wt% Ag/C (25.9 nm Ag particle size) as the electrocatalyst, the rechargeable ZABs had a power density of 34 mW cm^{-2} at 35 °C. Although the electrocatalytic performance was superior to Pt/C, the power density increased to 72 mW cm^{-2} at 80 °C, close to the operating temperature of the ZABs [3].

In another study, a novel gas diffusion layer (GDL) was designed using Ag-NPs supported by SWCNTs as the cathode of ZABs. The weight and thickness of the GDL of AgNP-SWCNTs were reduced to 0.005 mg cm^{-3} and 0.05 mm, respectively, and the electrode performance was comparable to that of the GDL of conventional carbon-based electrodes with weights and thicknesses of 0.79 mg cm^{-3} and 0.49 mm, respectively [4]. In addition, Ag-based catalysts possess better OER performance (i.e., more negative OER onset potential) in highly alkaline electrolytes compared to Pt/C. The poor OER performance of Pt/C is attributed to the accumulation of oxygenated substances which cover the surface and reduce the active surface area [5].

W. Huang et al. synthesized a porous N-doped carbon framework loaded with Ru NPs using Ru-doped ZIF-8 as a precursor and thermally treated at 900 °C under N_2 atmosphere [6]. In this work, the N-doped carbon texture not only improved the conductivity and durability but also provided the best platform to promote the uniform dispersion of Ru NPs. R. Zou et al. further prepared high-exposure Ru NPs supported on hierarchical porous carbon by using bimetallic MOFs as precursors [7]. Due to the similar crystalline structures of Ru-MOFs and Cu-MOFs, bimetallic MOFs consist of Ru and Cu nodes connected by BTC as organic ligands. As shown in Fig. 3.3a, during pyrolysis, Cu sites can act as spacers to isolate Ru sites and promote the formation of ultrafine and uniformly dispersed Ru NPs. Meanwhile, $FeCl_3$ solution can effectively remove Cu NPs, generating a large number of mesopores/macropores, facilitating charge transfer and forming highly exposed Ru NPs. Z. Chen et al. proposed a simple strategy to form ultra-small nanostructures encapsulated in a hollow-structured N-doped porous carbon matrix by using ZIF-8 as a sacrificial template material and tannic acid as a structural stabiliser Pt NPs (Fig. 3.3b) [8]. They first took advantage of the easy self-assembly and nucleation of ZIF-8 on Pt NPs to fabricate Pt NPs and ZIF-8 composites. Subsequently, tannic acid was coated on its surface to form a composite with a core–shell structure. In this work, tannin coating played a vital role in forming hollow porous carbon and highly distributed Pt NPs.

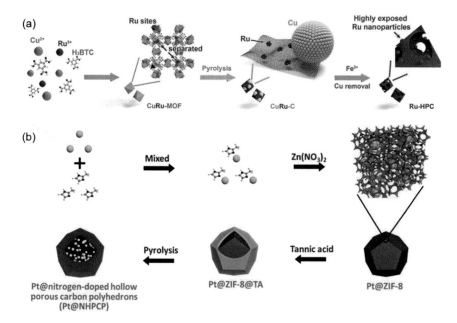

Fig. 3.3   a Formation of Ru NPs supported on hierarchical porous carbon [7]. **b** Schematic illustration of the fabrication of platinum NPs encapsulated in a hollow-structured N-doped porous carbon matrix [8]

3.2.3   Precious Metal Composite Non-precious Metal Based Catalysts

As an alternative to carbon, the combination of transition metal oxides with noble metal nanoparticles has been a hot research topic in recent years. Pt is one of the most expensive and least abundant metals. Other noble metals, such as silver (Ag), cost only 1% of Pt and are attractive due to their excellent stability and good mass activity. The required mass is minimized by alloying with low-cost metals/metal oxides with high ORR/OER activity, or nanostructured noble metals. The optimized electronic structure and increased active surface of these bimetallic catalysts have significantly increased activity and stability [9–12].

For example, Ag nanoparticle-modified $LaMnO_3$ chalcogenide nanorods/reduced graphene oxide (LMO-NR/rGO) were investigated as bifunctional catalysts. The catalysts were synthesized using a simple hydrothermal method. The Ag nanoparticles were uniformly dispersed onto the rGO flakes, improving the electrical conductivity while promoting rapid charge transfer [13] Combining noble metals with non-precious metal alloys such as Cu and Pb is also one strategy to reduce the cost and improve the long-term stability of ZABs [12, 14–16]. For example, the mass activity of structured Pd_3Pb/C catalysts is approximately two to four times higher than that of conventional fiducial Pt/C catalysts. The enhanced activity of Pd_3Pb/C results from more active sites generated by changing the Pd-Pd bond distance and altering the electronic configuration. The assembled ZABs exhibit favorable long-term stability, with an increase in overpotential of only 0.14 V after 560 h of cycling [16].

In another study, intermetallic structured CuPt-nanocages (CuPt-NCs) with a particle size of around 20 nm were prepared through a solvothermal method. The open-face hollow geometry of CuPt-NCs with ordered atomic arrangements provides easy O_2 molecular accessibility and many active sites within the structure. The application of CuPt-NCs in ZABs resulted in higher specific capacities and energy densities (560 mA h g_{Zn}^{-1} and 728 W h kg_{Zn}^{-1} with a discharge rate of 20 mA cm^{-2}) and Pt/C as cathodes (480 mA h g_{Zn}^{-1} and 624 W h kg_{Zn}^{-1} with a discharge rate of 20 mA cm^{-2}) [15]. Recently, dendritic Ag-Cu developed by current substitution reaction has been considered an active bifunctional catalyst. The advantages of this catalyst include the simple preparation process, the carbon-free structure and the high stability [17]. Furthermore, DFT analysis of a similar catalyst (Ag-Cu nanoalloys) prepared by pulsed layer deposition (PLD) showed an oxygen adsorption energy of -0.86 eV at pure Ag clusters and -1.36 eV at AgCu (Cu-shell) clusters. This type of catalyst promotes ORR by lowering the adsorption energy and facilitating the thermodynamic response of the ORR [12, 14].

As shown in Fig. 3.4, S. Mu et al. developed a general synthetic method to prepare Co, N-doped carbon substrates loaded with noble metal NPs (Ru, Pt, Pd, etc.) [18]. After immobilizing Ru NPs, the specific surface area of Co (single-atom catalysts) SACs decreased significantly. This may be because the ultrafine Ru NPs can be easily immersed into the micro/mesopores of Co SACs to improve the durability. The synergistic effect between metal NPs and SAs endows it with intriguing catalytic properties, offering potential for designing catalysts.

3.3 Transition Metal Bifunctional Catalysts

3.3.1 Transition Metal Oxides

It is well known that rechargeable zinc-air batteries generally use strong alkaline electrolytes such as 6 mol L^{-1} KOH, and non-noble metal oxides generally show good chemical stability in such electrolytes. Therefore, most of the current research results of bifunctional catalysts belong to this category and show good OER/ORR activity. The activity of partial transition metal oxide catalysts is comparable to that of noble metal-based composite catalysts. This section discusses the oxygen catalytic performance of metal oxide catalysts in detail according to their composition and structure.

Traditional catalysts (such as MnO_x and Ti_4O_7) have attracted extensive attention due to their simple components, environmental friendliness, low cost, etc. In terms of oxygen reduction catalysis, MnO_x can promote the disproportionation reaction of H_2O_2, which is beneficial for ORR to adopt the "4-electron" transfer path

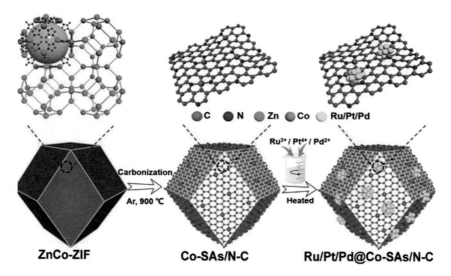

Fig. 3.4 A general synthetic method for the preparation of N-doped carbon-on-carbon co-doped metal clusters and SACs [18].

to realize the direct generation of hydroxide from oxygen molecules. In addition, another reaction mechanism of MnO_x has also been reported, that is, the presence of a large number of Mn^{3+}/Mn^{4+} redox-coupled ion pairs in MnO as acceptor–donor, which enables ORR to adopt a four-electron reaction pathway. Specifically, MnO_2 is first protonated to form MnOOH, then two MnOOH intermediates are combined with oxygen molecules to generate hydroxide. In OER catalysis, studies have shown that the valence state of manganese ions of MnO_x catalysts is related to OER activity. As far as the current research is concerned, high-valence Mn ions are more conducive to accelerating OER. The bifunctional catalytic activity of the catalyst is closely related to the structure of the metal oxide and the surface oxidation state species. On the other hand, J. He et al. directly pyrolyzed manganese oleate to obtain new nanostructured MnO_x such as nanoflocs, nanoflowers and octahedral nanoparticles [19], and its surface exposed different crystal plane structures. In this work, the structure–activity relationship between MnO_x, electrocatalytic activity and the interfacial structure of the material was studied in detail. The study showed that the nano-flocculated MnO_x had the highest ORR and OER activities, which were better than other morphologies of MnO_x. Although manganese oxides have been extensively studied, the key factors affecting ORR activity are not clear because it is not directly comparable. The techniques for manganese oxide synthesis are not the same in different studies, such as electrochemical deposition, hydrothermal methods, chemical methods, annealing, electrochemical treatments, etc. Each technique produces quite different crystals/particle sizes, shapes, porosity, and electronic conductivity. These properties not only affect the mass activity of the oxides but are also related to the electronic structure of the surface, resulting in different specific activities.

In non-precious metal catalysts, spinel oxides generally have ORR activity and OER properties. Spinel is generally represented in the form of AB_2O_4, in which the oxygen anions take the form of cubic close packing, 1/8 of the tetragonal points are occupied by A atoms, and 1/2 of the octagonal points are occupied by B atoms. A is generally a divalent metal ion (such as Mg, Fe, Co, Ni, Mn, Zn), and B is a trivalent metal ion (such as Al, Fe, Co, Cr, Mn). Spinel oxides with mixed valences exhibit conductive and semiconducting properties, which enable them to be used directly as electrode materials, and electron transfer can occur at relatively low activation energies. Transition metal oxides with spinel structures exhibit good ORR activity in alkaline media. Catalysts with different catalytic properties can be obtained by adjusting the composition of A^{2+} and B^{3+}. For example, J. Chen et al. discovered the rapid room-temperature synthesis of nanocrystalline $Mn_xCo_{3-x}O_4$ spinel and its application in catalytic zinc-air batteries. The most active $Mn_xCo_{3-x}O_4$ electrode can provide stable galvanostatic discharge voltage and considerable specific energy density at 1.3 V [20]. In recent years, J. Chen et al. have further developed phase-controllable spinel $CoMn_2O_4/C$ composites by oxidation-precipitation and crystallization methods. Nanosized $CoMn_2O_4$ exhibits excellent ORR/OER performance with a high energy density of 650 Wh kg^{-1} at 10 mA cm^{-2} and a voltage drop of only 8.5% after 155 cycles in a Zn-air battery. In addition, Z. Chen's group also designed pomegranate-shaped Co_3O_4 with abundant active sites, enhanced electron transfer and strong cooperative coupling. Then, a single-cell Zn-air battery was fabricated using this catalyst with almost no voltage drop during discharge or charge after 80 h of continuous operation [21].

Perovskite oxides have also received much attention as ORR catalysts in alkaline media. The perovskite ABO_3 includes a BO_6 octagonal structure shared at the corners and A cations at the corners. The perovskite structure with certain flexibility can withstand a certain amount of lattice mismatch between A-O and B-O, and accommodate a certain amount of auxiliaries between the A-site and the B-site. The substituted perovskite can be represented as $A_xA'_{1-x}ByB'_{1-y}O_3$, where A or A' is a rare earth element or an alkaline earth metal element, and B or B' is a transition metal element. This unique structure enables perovskites to have good ORR and OER properties and can be used as bifunctional catalysts. The oxygen reduction activity of perovskites is mainly related to the cations at the B-site and is weakly related to the cations at the A-site.

For example, A. Hammouche et al. prepared $La_{1-x}Ca_xCoO_3$ by sol–gel method [22] and found that when x = 0.4, the ORR rate and active specific surface area reached the maximum. J. Tolloch et al. studied the ORR catalytic behavior of $La_{1-x}Sr_xMnO_3$ (x = 0, 0.2, 0.4, 0.6, 0.8, and 1.0) in alkaline electrolytes [23] and found that $La_{0.4}Sr_{0.6}MnO_3$ had the highest catalytic activity, close to Pt black. Although there has been increasing progress in applying metal oxide-doped heteroatoms for ORR in alkaline fuel cells, the fundamental understanding of ORR mechanism is still lacking. Many possibilities for the composition of metal oxides make it challenging to reach a consensus on the relationship between material structure and catalytic performance. J. Suntivich et al. proposed an activity descriptor to predict the electrocatalytic activity of perovskite based on the electronic state factor of B-site doped metals [24]. Their study shows a volcano-shaped relationship between the filling state of the e_g orbital on the B-site of perovskite oxides and the ORR activity, as shown in Fig. 3.5. When the e_g orbital has moderate electron filling, the activity of oxides is the best, and the adsorption of oxygen-containing species at the B-site is neither strong nor weak, which is beneficial to the effective activation and desorption of oxygen-containing species. In general, with the increase of metal ions, the catalytic activity of perovskite oxides can be better regulated and often shows better performance. Still, its synthesis cost and preparation difficulty also increase. So far, ternary perovskite oxides are the most widely used in zinc-air batteries. In the future, with the further development of related processes, more metal-rich perovskite oxides and oxygen-hole-rich perovskite oxides will be reported more.

The research and development of transition metals as bifunctional catalysts has made significant progress. However, their catalytic activity is limited by the low electrical conductivity and aggregation of metal oxide nanoparticles. Studies have found that the problems of small specific surface area, poor conductivity and insufficient active sites of transition metal oxides can be improved by constructing nanostructures, compounding with carbon materials, or doping heteroatoms [25]. G. Cheng et al. supported Co_3O_4 nanoparticles on nitrogen-doped carbon materials (N-KC) by hydrothermal method [26]. The test showed the half-wave potential was 0.82 V, and the limiting current density in 0.1 mol L^{-1} KOH solution was 5.70 mA cm^{-2}. Due to the strong coupling effect of Co_3O_4 and N-KC and the synergistic effect of nitrogen element doping, the obtained Co_3O_4/N-KC nanocomposite exhibited remarkable catalytic activity. Y. Li et al. prepared Fe_3O_4 and nitrogen-doped

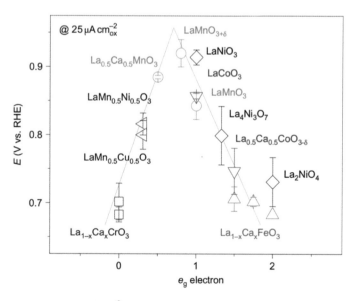

Fig. 3.5 Potential at 25 mA cm^{-2} in perovskite-based oxides as a function of the number of electrons in the e_g orbital [24]

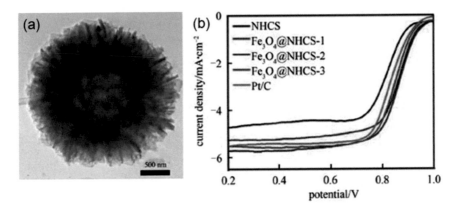

Fig. 3.6 **a** TEM image of Fe$_3$O$_4$@NHCS; **b** ORR polarization curve [27]

hollow carbon spheres composite catalysts (Fe$_3$O$_4$@NHCS) using ZnO nanowires as templates (Fig. 3.6a) [27]. The ORR half-wave potential and onset potential were better than those of commercial Pt/C catalysts (Fig. 3.6b). When used as a cathode electrocatalyst for Zn-air batteries, the power density and specific capacity are 133 mW cm^{-2} and 701 mAh g^{-1}, respectively. The analysis found that the hollow structure composed of high specific surface area and the synergistic effect of highly active nanoparticles and nitrogen-doped matrix materials were the main reasons for obtaining high-performance catalysts.

Mixed valence transition metal oxides (AB$_2$O$_4$) with spinel structure are typical composite metal oxides, which have more flexible structure control and better electrocatalytic performance than single metal oxides. They have excellent electrocatalytic performance and good durability in alkaline electrolytes. D. Lim et al. synthesized NiCo$_2$O$_4$ with a large number of oxygen vacancies by hydrothermal method and hydrogen reduction, with an onset potential and half-wave potential of 0.93 V and 0.78 V, respectively, and excellent durability and methanol resistance [28]. The analysis found that the improved catalyst activity and stability were attributed to the synergistic effect of the large electrochemical specific surface area, plenty of exposed active sites, and good electrical conductivity resulting from abundant oxygen vacancies. H. Ge et al. prepared a novel bifunctional electrocatalyst (NiCo$_2$O$_4$@NCNWs) composed of hollow NiCo$_2$O$_4$ nanospheres and N-doped carbon nanomesh through a liquid-phase synthesis method [29]. The N-doped carbon nanomesh mainly contributes to the high activity of the oxygen reduction reaction, while the NiCo$_2$O$_4$ nanospheres are the active phase for the oxygen evolution reaction. The Zn-air battery cathode catalyst exhibited low overpotential, achieved a specific capacity of 826.9 mAh g^{-1}, and only lost 9 mV of ORR half-wave potential after 5000 cycles, showing long-term stability.

Transition metal oxides (TMOs) (M=Fe, Co, Ni, Mn, etc.) derived from MOFs have attracted extensive attention as bifunctional electrocatalysts for ZABs. First, the variable valence and abundant structure of TMOs endow them with rich redox electrochemical properties. Second, the abundant organic ligands in MOFs enable the TMOs to be uniformly dispersed in the porous carbon matrix during the controlled carbonization process, which can effectively improve the conductivity and stability of the catalyst. Co-based MOFs derived from Co_3O_4 are considered as promising candidates for bifunctional electrode materials due to their high electrocatalytic activity among cobalt oxide materials. For example, C. Guan et al. synthesized a novel bifunctional catalyst (denoted as N-Co_3O_4/CC) by pyrolysis of Co-MOF, which consisted of N-doped carbon nanowall arrays embedded with hollow Co_3O_4 nanospheres on carbon cloth (Fig. 3.7a) [30]. The carbon onion coating suppressed the Kirkendall effect, promoted the conversion of Co nanoparticles into hollow Co_3O_4 nanospheres with n-granular polycrystalline oxide structure, and enhanced the bifunctional ORR/OER catalytic performance with an $E_{1/2}$ of ORR as high as 0.87 V (Fig. 3.7b). When assembled into a solid-state ZAB, it exhibited a high OCP of 1.44 V and a high-specific capacity of 387.2 mAh g^{-1}, surpassing platinum and iron-based devices. In addition, Y. Jiang et al. synthesized a hybrid composed of metallic Co and spinel Co_3O_4 nanoparticles embedded in a porous graphitized carbon shell via ion exchange and redox reactions between Co^{2+} and 2D Zn MOF nanosheets (termed Co/Co_3O_4@PGS) [31]. This hybrid catalyst possesses three distinct phases, metal, spinel oxide, and carbon, which are interconnected throughout the nanostructure. By constructing a three-phase structure of atom-trapping defect interfaces and porous graphitized carbon shells, charge transfer and utilization of catalyst active sites are significantly facilitated. Therefore, the Co/Co_3O_4@PGS catalyst exhibited significantly improved bifunctional catalytic activity with a ΔE of 0.69 V (Fig. 3.7c). When the hybrid catalyst was integrated into the air electrode of rechargeable ZABs, it showed exceptionally high cycling stability with no voltage decay for 800 h at 10 mA cm^{-2} (Fig. 3.7d). This unique interfacial engineering strategy will endow MOF-derived metal oxides with significantly improved bifunctional catalytic activity, excellent electrical conductivity, and high cycling stability.

Besides Co-MOFs, other metal-based MOFs, such as Mn-based MOFs, have also been selected to design carbon-coated metal oxide materials as efficient bifunctional electrocatalysts. For example, David Lou et al. prepared a bimetallic organic framework (Co/Mn-MIL-100) by hydrothermal method for the first time [32]. Due to the large difference in reduction potentials of cobalt and nickel, the Co/Mn-MIL-100-derived MnO/Co/PGC polyhedron composed of MnO/Co and porous graphitic carbon possesses abundant MnO/Co hybrid interfaces, rather than conventional bimetallic oxides (Fig. 3.7e). The resulting MnO/Co/PGC exhibits excellent bifunctional oxygen catalytic activity with a small ΔE of 0.82 V and long-term durability (over 20,000 s) due to the formation of desired components and heterostructures. Notably, the home-made ZAB with MnO/Co/PGC as the air cathode exhibited excellent battery performance, including a high peak power density of 172 mW cm^{-2}, a specific capacity of 872 mAh g^{-1} (Fig. 3.7f), and excellent cycling stability of without significant voltage decay (350 cycles, 116 h) (Fig. 3.7g), exceeds that of commercial ZAB based on Pt/C + RuO_2. This work highlights the synergistic role

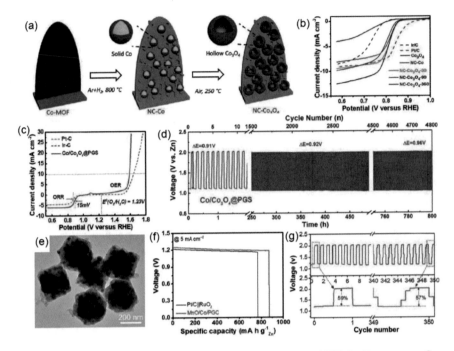

Fig. 3.7 **a** Schematic diagram of the fabrication process of layered NC-Co$_3$O$_4$ arrays on flexible carbon cloth. **b** Oxygen reduction polarization curve at a rotational speed of 1600 rpm [30]. **c** Whole LSV curves of bifunctional activities within the ORR and OER potential windows of Co/Co$_3$O$_4$@PGS, Pt/C and Ir/C at 1600 rpm. **d** Co/Co$_3$O$_4$@PGS catalyst at a current density of 10 mA cm^{-2} [31]. **e** TEM image of MnO/Co/PGC. **f** Discharge curves of ZABs at 5 mA cm^{-2}. **g** Long-term cycling performance at a current density of 10 mA cm^{-2} with MnO/Co/PGC as the air cathode [32]

of heteroplanes in oxygen electrocatalysis, thereby providing a promising avenue for advanced zinc-air cathode catalysts.

Transition metal oxides (TMOs) are one of the important families of bifunctional oxygen electrocatalysts for ORR and OER, of which spinel oxides, perovskites, and rutile-type oxides are outstanding representatives [33–35]. Compared with noble metal electrocatalysts, ideal activity, high abundance, low cost, and environmental friendliness are their fundamental advantages. Although some of them are unstable in acidic electrolytes, most are generally stable in alkaline media, making these TMOs practical as bifunctional oxygen electrocatalysts for fuel cells and metal-air batteries. This section will systematically review the electrocatalytic properties, influencing factors, and design strategies of various spinel oxides, perovskite, and rutile-type oxides, respectively.

Spinel oxides are receiving extensive attention as bifunctional ORR/OER electrocatalysts. A typical spinel oxide can be described as a group of oxides with the formula AB$_2$O$_4$, where A is a divalent metal ion (e.g., Zn, Cu, Mg, Fe, Co, Ni, Mn) and B is a trivalent metal ion (such as Al, Fe, Co, Cr or Mn). Typically, A^{2+} occupies part or all of the center positions of a tetrahedron, B^{3+} occupies part or all of the center position of an octahedron, and O^{2-} is located at the vertices of a polyhedron [36, 37]. The distribution of A^{2+} and B^{3+} is complicated. According to their different dispersion ratios in the octahedral and tetrahedral interstitials, spinel oxides can be classified into three types: ortho-spinel, anti-spinel and composite spinel, as shown in Fig. 3.8a. Three main factors affect the distribution of A^{2+} and B^{3+}, including the radius of the cation, the coulomb interaction between the cations and the crystal field of the cation's octahedral site preference energy [38]. In addition, some external factors can also cause structural changes, such as $CoMn_3O_4$ at high temperature or high pressure. Due to the redistribution of Co^{2+} and Mn^{3+}, a reversible structural transition occurs, that is, normal ↔ opposite [20]. As for spinel oxides, controllable structure, precise composition, and desired morphology (micro/nanostructure) are some alternative strategies to optimize their ORR/OER bifunctional electrocatalytic activity.

J. Chen et al. have done extensive work on the Co-Mn-O spinel system. For example, they successfully synthesized cubic and tetragonal CoMnO spinels through an oxidation-precipitation process at 180 °C, as shown in Fig. 3.8b. Compared with tetragonal, cubic CoMnO spinel has higher intrinsic ORR activity but poorer OER activity due to stronger oxygen binding capacity [39]. It has also been reported

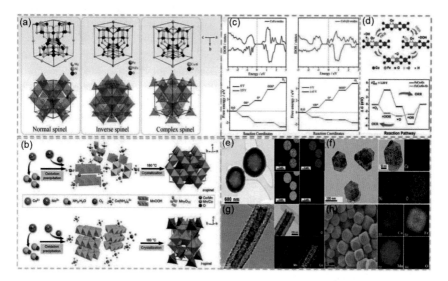

Fig. 3.8 **a** Representative structures of normal spinel ($MgAl_2O_4$), inverse spinel ($NiFe_2O_4$) and composite spinel ($CuAl_2O_4$). The green and purple polyhedra correspond to the metal occupancy of octahedra and tetrahedra, respectively [38]. **b** Schematic diagram of the synthesis of cubic and tetragonal spinel phases. **c** The density of states and free energy diagrams of pure CoFe oxide and CoFeZr oxide. **d** OER/ORR mechanism of R-$FeCo_2O_4$ with oxygen vacancies and DFT plots of $FeCo_2O_4$ and R-$FeCo_2O_4$ [55]. **e** TEM and elemental map images of $ZnMnCoO_4$ microspheres [76]. **f** $NiCo_2O_4$ nanodisks [69]. **g** $MnCo_2O_4$ hollow nanofibers. **h** $Mn_{0.6}Fe_{1.2}Co_{1.2}O_4$ nano-dice [71]

that Fe-doping can adjust the phase transition of Co_3O_4 spinel, and it was found that normal Co_3O_4 spinel could be adjusted to its reverse phase by controlling the content of Fe-doping and then back to the normal phase due to the elongation O–O bond and modulated oxygen adsorption energy, the reversed-phase with the $(Co)[FeCo]O_4$ form showed the highest ORR activity [40]. Biphasic $MnCo_2O_4$ (dp-MCO) with cubic and tetragonal spinels was prepared by a simple hydrothermal reaction and post-lyophilization treatment. Compared with the commercial Pt/C (30 wt% Pt) electrocatalyst, the as-prepared dp-MCO performed well due to the optimized Mn^{4+}/Mn^{3+} redox pair (ORR active site) and the increase of surface Co^{3+} cations (OER active site). The as-prepared dp-MCO demonstrated excellent ORR/OER bifunctional activity, enabling the rechargeable Zn-air battery to perform 64 discharge-charges ($\approx$768 h) successfully [41].

Insights into the relationship between the crystal planes of spinel oxides and catalytic activity play a crucial role in designing efficient bifunctional oxygen electrocatalysts. Through a simple template-free hydrothermal strategy, Han controllably synthesized Co_3O_4 nanocubes, nanooctahedrons, and nanopolyhedra with exposed crystal faces (001), (001) + (111), and (112), respectively. Their study showed that the adsorption, activation, and desorption of oxygen-containing species were optimized here with abundant octahedral-coordinated Co^{3+} sites on the (112) plane and benefited ORR/OER activity and Zn-air batteries performance [42]. Similar work was done by Q. Liu et al. on the exposed planes of (100), (110) and (111), respectively. They believed that the surface Co^{3+} cations were beneficial for higher ORR performance [43].

Incorporating other elements is an effective way to adjust the electrical structure, change the chemical composition and valence state of spinel oxides, and improve their electrocatalytic performance [44]. Alkali metals, transition metals, rare earth metals and even non-metallic elements can be selected as doping elements [45–48]. J. Liu et al. reported Li-doped Co_3O_4 solid solution nanocrystals as ORR/OER electrocatalysts. They found that Co_3O_4 doped with 5% Li exhibited the best activity, and the incorporation of Li not only helped to increase the content of covalent O=C–O–Co–III–O bonds greatly, but also promoted the oxidation of Co(III) to Co(IV), but also has a significant benefit in promoting hydroxide substitution on the surface [49]. The effect of Ni incorporation on the bifunctional ORR/OER activity of $MnCo_2O_4$ was discussed by Schechter et al. Their study showed that the optimized $Ni_{1.5}Co_{0.75}Mn_{0.75}O_4$ catalyst presented the best ORR/ORR activity with an overpotential gap ($\Delta E = E_{OER} - E_{ORR}$) of 0.79 V versus RHE between the two reactions in alkaline solution. The substitution of nickel leads to the in situ formation of ternary phases in the catalyst, which is a crucial factor in improving the activity and durability [50]. It has also been reported that iron incorporation into cobalt vanadate (CoV_2O_4) transformed CoV_2O_4 from a normal $CoFe_2O_4$ spinel to a reversed $CoFe_2O_4$, and $CoV_{1.5}Fe_{0.5}O_4$ exhibited a small ΔE of 0.83 V versus RHE in 1.0 M KOH solution [51]. Recently, it was found that adding Zr to $CoFe_2O_4$ could tune the local electronic environment of spinel and then induce charge transfer between Fe and Co ions. Meanwhile, the oxygen adsorption performance of the oxide surface can also be optimized (Fig. 3.8c). Therefore, the as-prepared 3D CoFeZr oxide nanosheets exhibited excellent OER activity with a low overpotential of 248 mV at 10 mA cm^{-2} in 1.0 M KOH [52]. In addition, their group reported the treatment of C-doped Co_3O_4 by CH_4 plasma and found that a large number of oxygen vacancies and new CoC bonds (C doping) could be achieved, which would tune the Co valence distribution in Co_3O_4 and improve its oxygen electrocatalytic activity.

Oxygen vacancies are the most popular anionic defects, and many researchers believe that oxygen vacancies can tune the surface electronic properties and interstitial states of spinel oxides, which is beneficial to improving oxygen catalytic activity. For example, J. Sun et al. used ionic liquids as dopants to introduce oxygen vacancies into $CoFe_2O_4$ spinel, showing the best OER activity for spinel-based electrocatalysts with an overpotential of 250 mV and a small Tafel slope of 41 mV dec^{-1} [53]. A new "adsorption-calcination-reduction" strategy was developed by S. Peng et al. to prepare a series of spinel oxides ($NiCo_2O_4$, $CoMn_2O_4$, $NiMn_2O_4$) with unique yolk-shell structure and abundant oxygen vacancies, which have excellent OER activity [54]. Kang-Wen et al. [55] prepared typical $FeCo_2O_4$ spinel nanoparticles with abundant oxygen vacancies by laser fragmentation, which exhibited excellent bifunctional oxygen electrocatalytic performance with a low half-wave potential of 0.82 V for ORR and required a minimum overpotential of 276 mV to sustain 10 mA cm^{-2} for OER. As shown in Fig. 3.8d, DFT calculations showed that oxygen vacancies could effectively lower the thermodynamic energy barrier and accelerate electron transfer. S. Wang et al. reported that Co_3O_4 nanosheets require only an OER potential of 1.53 V versus RHE after Ar plasma etching to reach a current density of 10 mA cm^{-2}, a potential reduction of 240 mV compared to pristine Co_3O_4 nanosheets. Abundant oxygen vacancies and increased surface area lead to enhanced OER performance [56].

As for spinel bifunctional oxygen electrocatalysts, much work has focused on designing micro/nanostructures with various morphologies, including quantum dots [57, 58], nanoparticles [59, 60], nanorods [61, 62], nanowires [63, 64], nanotubes [65, 66], nanofibers [67, 68], nanoflakes [69, 70], nanocubes [71–73], nanoflowers [74, 75], hollow microspheres [76, 77], and nanowire arrays [78, 79] to enhance active sites and improve catalytic activity (Fig. 3.8e–h). $CoMn_2O_4$ quantum dots (CMO QDs) were prepared by a combined thermal injection and heating method. CMO QDs with an average size of 3.9 nm possessed low bandgap energy, moderate surface oxygen adsorption energy and large charge carrier concentration, exhibiting the highest ORR/OER activity among all investigated materials [80]. $CuCo_2O_4$ QDs, $NiFe_2O_4$ QDs, and $CoMn_2O_4$ nanodots have also exhibited desirable ORR/OER performance due to the high proportion of active edge dots [81–83]. Recently, K.-N. Jung et al. fabricated 1D $MnCo_2O_4$ and $CoMn_2O_4$ nanorubbers by electrospinning technology as ORR/OER bifunctional electrocatalysts for rechargeable Zn-air batteries, which can achieve reduced discharge–charge voltage gap [84]. The mesoporous thin-walled $CuCo_2O_4$@C hollow nanotubes were prepared by the same electrospinning process, which significantly reduced the charge–discharge voltage gap (0.79 V at 10 mA cm^{-2}) and extended the long cycle life of Zn-air batteries (up to 160 times within 80 h) [85]. The reported nanoengineered core–shell $NiCo_2O_4$ chestnut-like structure exhibited a small ORR/OER potential difference of 0.90 V in 0.1 M

KOH solution [86]. C. Jin et al. developed a facile template-free co-precipitation route to fabricate ordered $NiCo_2O_4$ (NCO) spinel nanowire arrays, which avoided collapse during fabrication and possessed specific surface areas as high as 124 m^2 g^{-1} and provided ORR/OER bifunctional activity close to commercial Pt/C and IrO_2 electrocatalysts [87]. Through using PS colloidal microspheres and KIT-6 mesoporous silica as hard templates, three-dimensional ordered macroporous (3DOM) $CoFe_2O_4$ and $CuCo_2O_4$ were prepared as bifunctional oxygen electrocatalysts [88, 89].

Considering the relatively low conductivity of spinel oxides, modifications with high conductive materials have been carried out. X. Cao et al. fabricated conductive polypyrrole-modified $MnCo_2O_4$ spinel as ORR/OER electrocatalyst [90]. Carbon-coated $MnCo_2O_4$ (MCO) nanowires have also been reported as bifunctional oxygen catalysts for rechargeable zinc-air batteries. The carbon shell on the surface not only provides a conductive network to facilitate electron transfer but also restricts the growth of MCO nanoparticles during discharge–charge [91]. In addition, Han et al. [92] constructed a conductive MnCoP shell in situ on the MCO spinel surface by a facile phosphating method, requiring only an OER overpotential of 269 mV to achieve a current density of 10 mA cm^{-2}. Furthermore, X. Cao et al. [93] reported a bifunctional electrocatalyst decorated with $MnCo_2O_4$ and highly conductive Ti_4O_7, which was synthesized by a facile solution thermal method. The synergistic effect between Ti_4O_7 and $MnCo_2O_4$ improved the ORR/OER kinetics.

Several other potential factors can affect the ORR/OER properties of spinel. For example, X. Cao et al. reported that the calcination temperature affected the size and porosity of $MnCo_2O_4$, which then had a non-negligible effect on its bifunctional catalytic activity [94]. Furthermore, J. Zhao et al. developed a strategy to engineer active sites to enhance the ORR/OER activity of $NiCo_2O_4$ (NCO) spinel by simply adjusting the annealing temperature. They found that NCO-250 (250 °C) showed excellent bifunctional oxygen electrocatalytic performance due to the highest value of Ni^{3+}/Ni^{2+} sites and the lowest value of Co^{3+}/Co^{2+} sites [95]. The ORR/OER performance of some recently reported sipnel oxides are summarized in Table 3.1.

Perovskites are promising non-orthogonal oxide functional materials that can be described by the general formula $ABO_{3-\delta}$ (Fig. 3.9a), where A is a rare earth and/or alkaline earth element (La, Pr, Sr, Ca, Ba), B is a transition metal element (Fe, Co, Ni, Mn, Cr, Cu) [96–98]. Most of the $ABO_{3-\delta}$ perovskites are mixed ion–electron conductors due to the abundant oxygen vacancies and B elements of various valences, which endow them with excellent oxygen electrocatalytic activity. In this section, we briefly review feasible strategies for optimizing the catalytic activity and stability of perovskite.

The electrocatalytic properties of perovskite can vary dramatically with different A-site or B-site cations. Generally speaking, the substitution of the A site by alkaline earth elements mainly affects the ability to adsorb oxygen and favors the generation of oxygen vacancies, while the doping of the B site by transition metal elements affects the activity of adsorbing oxygen and helps to tune the electron conductivity [99, 100]. For example, Y. Zhao et al. discussed the effect of Sr doping on the bifunctional oxygen catalytic activity of $LaMnO_3$ electrocatalysts [101]. They found that $La_{0.4}Sr_{0.6}MnO_3$ not only exhibited higher surface adsorbed oxygen content and

Table 3.1 Summarising ORR/OER bifunctional performance of the aforementioned spinel oxide-based electrocatalysts

Catalysts	$E_{1/2}$ (V)	η_{10} (V)	$E_{10} - E_{1/2}$ (V)
$MnCo_2O_4$ [323]	0.83	0.4	0.8
$NiCo_2O_4$ [324]	0.73	0.4	0.9
$CoMn_2O_4$ [325]	0.75	0.6	1.08
$Co@Co_3O_4$ [326]	0.78	0.41	0.87
$NiCo_2O_4$ [327]	0.82	0.35	0.76
$CuCo_2O_4$ [328]	0.8	0.47	0.9
$ZnCo_2O_4$ [329]	0.8	0.42	0.85
$NiCo_2O_4$ [330]	0.65	0.41	0.99
Fe-doped CoV_2O_4 [331]	0.66	0.3	0.83
Ni-doped Co_3O_4 [332]	0.83	0.38	0.78
$NiCo_2O_4$ [333]	0.63	0.42	1.02
$d\text{-}Co_3O_4$ [334]	0.88	0.27	0.62
$O_v\text{-}NiCo_2O_4$ [335]	0.74	0.44	0.93
$O_v\text{-}FeCo_2O_4$ [336]	0.82	0.28	0.69
$O_v\text{-}Co_3O_4$ [337]	0.87	0.42	0.78
$MnCo_2O_4$ [338]	0.72	0.51	1.02
$Co_3O_4/N\text{-}rGO$ [339]	0.76	0.38	0.85
$NiMnO_3/NiMn_2O_4$ [340]	0.75	0.38	0.86
$MnCo_2O_4/C$ [341]	0.77	0.43	0.89

larger specific surface area but also increased the content ratio of Mn^{4+}/Mn^{3+}. The Sr-doped $PrBa_{1-x}Sr_xCo_2O_{5+\delta}$ (x = 0–1.0) double perovskite is also reported here. On the one hand, the Sr substitution increases the covalency between the Co 3d and O 2p orbitals due to lattice shrinkage. On the other hand, Sr doping disrupts the ordering of the PrO and Ba(Sr)O layers, increases the concentration of Co^{4+} (g = 1), promotes the formation and deprotonation of OOH species, and the adsorption of O_2. These factors contribute to its remarkable ORR/OER electrocatalytic behavior [102]. As for $La_{1-x}M_xCoO_3$ (M=Ca, Sr), Ca-incorporated materials showed slightly better ORR performance, while Sr-substituted materials showed better OER activity [103]. The substitution of Fe to $La_{0.6}Ca_{0.4}CoO_{3-\delta}$ was also discussed, suggesting that the oxidation of Fe^{3+} to Fe^{4+} and the partial reduction of Co^{3+} to Co^{2+} might be the reason for the enhanced ORR/OER activity [104]. In addition, doping of non-metallic elements positively affected the bifunctional oxygen catalytic activities of perovskites, including P-, S-, and F-doping [105–109]. For example, S. Peng et al. controlled the large-scale preparation of sulfur-doped $CaMnO_3$ nanotubes as ORR/OER electrocatalysts for Zn-air batteries by electrospinning technology.

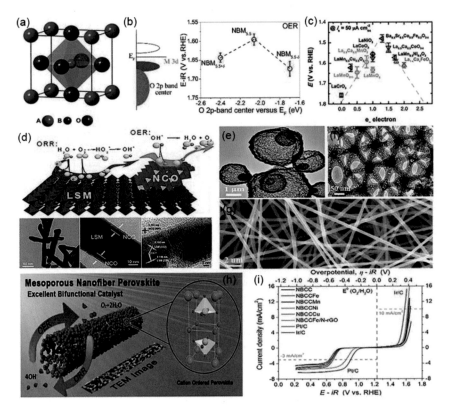

Fig. 3.9 **a** Crystal structure of ABO$_{3-\delta}$ cubic perovskite. **b** Schematic diagram of the O 2p band center and the dependence of the OER activity of perovskite on the O $2p$ band center and the Fermi level [110]. **c** The OER activity (potential @50 μA cm^{-2}) of some perovskite is shown by a volcano-type relationship as a filling function of transition metal B ions in ABO$_3$ [111]. **d** Schematic diagram of the ORR and OER reaction mechanisms and TEM image of NiCo$_2$O$_4$ @La$_{0.8}$Sr$_{0.2}$MnO$_3$ [112]. **e** TEM image of La$_{0.9}$Sr$_{0.1}$CoO$_3$ perovskite microspheres with egg yolk-shell structure [113]. **f** HR-TEM image of 3D ordered macroporous LaFeO$_3$ [114]. **g** SEM image of porous La$_{0.75}$Sr$_{0.25}$MnO$_3$ nanotubes [115]. **h** Various cationically ordered PrBa$_{0.5}$Sr$_{0.5}$Co$_{2-x}$Fe$_x$O$_{5+\delta}$ hollow nanofibers prepared by electrospinning [116]. **i** Overall oxygen electrocatalytic performance of different samples [117]

It is experimentally demonstrated that the S doping engineering increased the electrical conductivity and the surface vacancy defects, thereby enhancing the adsorption capacity of O$_2$, which greatly improved the performance of the Zn-air battery.

Similar to the Pt, IrO$_2$, and RuO$_2$ electrocatalysts, the crystallographic orientation of the peridotite oxides also exerts a significant influence on the catalytic activity of bifunctional oxygen. For example, the (001) orientation of La$_{0.6}$Ca$_{0.4}$CoO$_3$ showed higher ORR/OER performance than the (110) orientation [118]. It was found that (001)-oriented LaMnO$_3$ single-crystal flakes showed intrinsic ORR activity comparable to LaMnO$_3$ polycrystals, with high surface area, more oxygen vacancies and Mn^{3+} active sites for ORR [119]. In addition, different planes of SrRuO$_3$ exhibited various ORR activities in decreasing order: $SrRuO_3$ (111) > $SrRuO_3$ (110) > $SrRuO_3$ (001), and the density of surface defects controlled their OER performance.

The occupancy of the O $2p$ band centre relative to the Fermi energy level (a measure of the ionic bond strength of the perovskite to the intermediate) and the e_g orbital (a measure of the strength of the covalent bond) are the two main factors that are good descriptors of the ORR as well as the OER on various perovskite catalysts. Compared with disordered $Nd_{0.5}Ba_{0.5}MnO_3$ and $NdBaMn_2O_{5.5}$ and $NdBaMn_2O_{5.5+\delta}$ ($\delta < 0.5$), double perovskites have stronger bifunctional activity, as shown in Fig. 3.9b [110]. According to the molecular orbital principle and e_g orbital filling of B ions, Suntivich et al. found that the intrinsic ORR and OER activities of perovskites exhibited volcano-like trends, respectively. When the e_g orbital filling degree is close to 1.0, $LaMnO_{3+d}$ exhibits the best intrinsic ORR activity, while $Ba_{0.5}Sr_{0.5}Co_{0.8}Fe_{0.2}O_3$- (BSCF) exhibits the highest intrinsic OER activity. As shown in Fig. 3.9c, they are comparable to the state-of-the-art Pt/C electrocatalyst for ORR and the IrO_2 electrocatalyst for OER.

These two perovskites also have some disadvantages. As for $LaMnO_{3+\delta}$, theoretical calculations and experimental results suggest that, although it exhibits the highest intrinsic ORR activity, it is not an optical OER catalyst. Y. Luo et al. coated the surface of $La_{0.8}Sr_{0.2}MnO_3$ (LSM) nanorods with high ORR-active nanolayers ($NiCo_2O_4$ shells) to obtain core–shell NCO@LSM heterojunctions as ORR/OER electrocatalyst, as shown in Fig. 3.9d [112]. Low electronic conductivity and large ORR overpotential are the big drawbacks of BSCF, which limit its low-temperature applications [120, 121]. The addition of conductive carbon to fabricate BSCF composite electrocatalysts is an effective method to tune its ORR/OER performance [122]. C. Jin et al. believed that the electronic interaction between carbon and BSCF was the driving force for the enhanced activity [123]. J. Il Jung et al. synthesized ($(La_{0.3}(Ba_{0.5}Sr_{0.5})_{0.7}Co_{0.8}Fe_{0.2}O_{3-\delta})$) as an ORR/OER bifunctional catalyst, and surprisingly, they discovered a novel structure, namely the inner of the cubic BSCF was surface-dispersed with rhombohedral $LaCoO_3$ nanoparticles (≈ 10 nm), which had optimized ORR/OER activity comparable to the state-of-the-art RuO_2 and IrO_2 catalysts [121]. Furthermore, a BSCF-decorated (001)-oriented LSM flake electrode is reported as a bifunctional catalyst, which exhibited ORR/OER activity that suppressed the activity of single LSM and BSCF.

Although most perovskites show relatively high intrinsic ORR and/or OER activity (specific surface area activity), high temperature (> 800 °C) calcination and long annealing processes must be required to synthesize crystalline perovskites, which leads to low specific surface area and mass activity. Therefore, there is an urgent need to develop new synthetic methods to prepare nanostructured materials with various morphologies, including nanoparticles, hollow spheres, sea urchin-like microspheres, hollow nanotubes, hollow nanorubbers, yolk-shell-structured microspheres, etc. [116, 124–127]. For example, S. Bie et al. prepared $La_{0.9}Sr_{0.1}CoO_{3-\delta}$ perovskite microspheres with egg yolk-shell structure by a one-pot hydrothermal method, as shown in Fig. 3.9e, and the number of shells can be controlled by using different solvents [113]. Three-dimensional ordered macroporous $LaFeO_3$ (3DOM-$LaFeO_3$) was synthesized by a facile colloidal crystal template method and used as a bifunctional oxygen catalyst. Compared with $LaFeO_3$ nanoparticles, 3DOM-$LaFeO_3$ exhibited good electrocatalytic activity and stability (Fig. 3.9f) [114]. Furthermore, it has been reported that hollow porous $La_{0.75}Sr_{0.25}MnO_3$ nanotubes (Fig. 3.9g) fabricated by electrospinning can be used as superior ORR/OER catalysts [115].

Besides $ABO_{3-\delta}$ cubic perovskites, some $A_2BO_{4+\delta}$ layered and $AA'B_2O_6$ bimorphic peroxides have been reported as ORR/OER bifunctional electrocatalysts. ORR onset potential and half-wave potential and OER activity showed the same trend, increasing sequentially, ie. $La_2NiO_4 < La_{1.9}Ca_{0.1}NiO_4 < La_{1.9}Sr_{0.1}NiO_4 < La_{1.7}Sr_{0.3}NiO_4$ [128]. To overcome the inherent low electronic conductivity of $La_{0.6}Sr_{1.4}MnO_{4+\delta}$, Wang et al. [129] fabricated a carbon coating layer on La_2NiO_4. Through a bottom-up strategy, a carbon coating with a thickness of less than 5 nm was fabricated on the surface of $La_{0.6}Sr_{1.4}MnO_{4+\delta}$, and its ORR activity and stability were improved. $Co_{1.5}Fe_{0.5}O_{5+\delta}$ (PBSCF) double peroxide was reported as an ORR/OER bifunctional electrocatalyst, and the porous PBSCF nanorubber prepared by the electrospinning method showed 20 times higher mass activity and ≈ 1.6 fold the intrinsic activity [130]. Y. Bu et al. also prepared mesoporous $PrBa_{0.5}Sr_{0.5}Co_2$-$FeO_{5+\delta}$ ($\delta = 0$, 0.5, 1, 1.5, and 2) peroxide nanorubbers, as shown in Fig. 3.9h, which exhibited good B-site cations. The ordered structure and large surface area (≈ 20 m^2 g^{-1}) ultimately lead to the excellent electrochemical performance of Zn-air batteries. Transition metal oxides (Fe^{2+}, Ni^{2+}, Cu^{2+}, and Mn^{2+}) were doped into $NdBa_{0.75}Ca_{0.25}Co_2O_{5+\delta}$ double peroxide catalysts, among which Fe-doped oxides exhibited the best ΔE (overpotential difference between ORR and OER at selected current densities) value of 0.948 V, as shown in Fig. 3.9i [117]. Some scholars have studied $Sr_{2-x}La_xFeMoO_6$ (x = 0, 0.25, 0.5, 1.0) and Co-doped Sr_2FeMoO_6 as bifunctional oxygen catalysts. High electrical conductivity contributes to enhancing the electrocatalytic activity of ORR and OER [131, 132]. Some new peroxides have also been studied. For example, $Ba_{0.9}Co_{0.5}Fe_{0.4}Nb_{0.1}O_3$ showed excellent bifunctional activity [133]. $CaMnO_3$ peroxides with open tunnels and multivalents showed excellent ORR/OER activity [134].

Recently, some new strategies have emerged to optimize the ORR/OER bifunctional electrocatalytic activity. Z. Wang et al. prepared La_2O_3-NCNTs in situ by forming well-dispersed metal and/or alloy nanoparticles (NPs) on the surface of the parent peroxide due to the precipitation of B cation in ABO_3 peroxide in a reducing atmosphere. As a new powerful bifunctional oxygen electrocatalyst, the hybrid derived from $LaNi_{0.9}Fe_{0.1}O_3$ peroxide has the typical morphology of nitrogen-doped carbon nanotubes. The coupling between NCNTs and La_2O_3 leads to the formation of active LaO and CO bonds, resulting in an excellent bifunctional activity [135]. However, $LaNi_{0.9}Fe_{0.1}O_3$ was decomposed in this work. In further work, Y. Wang et al. chose $La_{0.8}Sr_{0.2}Cr_{0.69}Ni_{0.31}O_3$-(LSCN) as the parent peroxide and obtained the LSCN/Ni_2P heterostructure double via a simple reduction and in situ phosphating process. The island-shaped Ni_2P nanoparticles are well embedded into the surface of the LSCN framework. Due to the unique LSCN/Ni_2P interface, excellent OH^- adsorption enhanced electrical conductivity and optimized O 2p band center position, the OER activity is increased by ≈ 6.2 times, and the mass activity is increased by ≈ 10.2 times [136]. Likewise, a layered peroxide with CoP nested in anoxic

Table 3.2 The summarization of the ORR/OER bifunctional electrocatalytic parameters of some relevant perovskites

Catalysts	$E_{1/2}$ (V)	η_{10} (V)	$E_{10} - E_{1/2}$ (V)
$La_{0.9}Sr_{0.1}CoO_3$ 113	0.68	0.52	1.07
$PrBa_{0.5}Sr_{0.5}Co_{2-x}Fe_xO_{5+\delta}$ [342]	0.69	0.3	0.84
$La_{0.5}Sr_{0.5}CoO_{2.91}$ [343]	0.76	0.6	1.07
$La_{0.7}(Ba_{0.5}Sr_{0.5})_{0.3}Co_{0.8}Fe_{0.2}O_{3-\delta}$ [344]	0.66	0.37	0.94
$La_{0.5}Sr_{0.5}CoO_{3-x}$ [345]	0.66	0.4	0.97
N-doped $LaNiO_3$ [346]	0.67	0.53	1.09
P-doped $SrCo_{0.5}Mo_{0.5}O_3$ [347]	0.75	0.45	0.93
S-doped $CaMnO_3$ [348]	0.76	0.47	0.94
$PrBa_{1-x}Sr_xCo_2O_5$ [349]	0.7	0.42	0.95
$La_{1.7}Sr_{0.3}Co_{0.5}Ni_{0.5}O_{4+\delta}$ [350]	0.57	0.59	1.25
$PrBa_{0.85}Ca_{0.15}MnFeO_{5+\delta}$ [351]	0.77	0.4	0.86
$Ba_{0.5}Sr_{0.5}Co_{0.8}Fe_{0.2}O_3$ [352]	0.77	0.35	0.81

$PrBa_{0.5}Sr_{0.5}Co_{1.5}Fe_{0.5}O_5$, prepared by in situ dissolution and an attractive "post-growth" approach, offered excellent power density (138.0 mW cm^{-2}), low charge–discharge voltage gap, and excellent stability in zinc-air batteries, and exhibited a competitive overpotential (0.440 V) in bulk water separation [137]. Some reported ORR/OER bifunctional perovskite electrocatalysts are listed and compared in Table 3.2.

In this section, the recent progress of other typical ORR/OER bifunctional electrocatalysts based on transition metal oxides, such as those related to CeO_2, manganese oxide, nickel oxide, and other oxides, is briefly summarized.

CeO_2 with fluorite cubic structure is an important family of functional ceramics. Ce^{4+}/Ce^{3+} redox couples, Ce^{3+} and abundant surface oxygen vacancies are useful active sites for ORR/OER [138]. There is a lot of literature on CeO_2 as a carrier material to disperse Pt nanoparticles [139–141]. CeO_2 not only inhibits Pt oxidation due to the oxidation of Ce^{3+} to Ce^{4+} but also prevents the growth of Pt nanoparticles, ultimately leading to enhanced ORR activity and stability. C. Jin et al. synthesized gold nanoparticle-decorated $Gd_{0.3}Ce_{0.7}O_{1.9}$ nanotubes by a simple self-etching method, achieving better ORR performance [142]. Doped CeO_2 has also been reported as a bifunctional oxygen catalyst for metal-air batteries [143, 144]. For example, R. S. Kalubarme et al. showed that a good discharge capacity of 8435 mAh g^{-1} and a lower discharge–charge potential gap could be obtained by utilizing Zr-doped CeO_2 as an electrocatalyst [145]. Y. Zhu et al. prepared MnO_x-decorated CeO_2 nanorods as efficient ORR/OER electrocatalysts [146]. F. Song et al. reported a $FeCo_2O_4/CeO_2$ heterostructure as an ORR/OER bifunctional electrocatalyst in 0.1 M KOH, which exhibited the lowest ORR/OER overvoltage gap of 1.009 V versus RHE. The synergy and the interface between $FeCo_2O_4$ and CeO_2 contributed to the excellent bifunctional activity [147].

Manganese oxides have attracted attention as bifunctional electrocatalysts for oxygen electrochemistry due to their different structures and valences. The crystal structure plays an essential role in the catalytic activity of manganese dioxide in the order $\alpha \rightarrow \beta \rightarrow \gamma$-manganese dioxide [148]. T. Ohsak et al. believed that chemical composition was also a critical factor in affecting the electrocatalytic performance of manganese oxide [149], and the trend are $MnOOH > Mn_2O_3 > Mn_3O_4 > Mn_5O_8$. Incorporating low-valent metal elements (Ni, Mg, Ca, etc.) can promote the transfer of O_2 and ads, stabilize the Mn^{3+}/Mn^{4+} redox couple in manganese oxide materials, and thus enhance the activity of ORR/OER. M. Liu et al. fabricated amorphous MnO nanowires on conductive KB carbon frameworks as electrocatalysts for rechargeable Zn-air batteries with a maximum power density of ≈ 190 mW cm^{-2} [150]. Theoretically, NiO is intrinsically inactive because its NiO binding strength is too strong or weak. Incorporating iron and/or cobalt can optimize its ORR/OER activity to some extent. Ni_1-FeO has been widely studied as a bifunctional oxygen catalyst, and $Ni_{0.9}Fe_{0.1}O$ exhibits similar ORR activity to BSCF [151]. The formation of the $Ni_{0.9}Fe_{0.1}OOH$ layered double hydroxide structure and the partial charge transfer activation effect on Ni after Fe doping may improve performance. In addition, Co-doped NiO has also been reported as an oxygen electrocatalyst, but its function in NiO has not been fully understood [152].

Considering the commercial utilization of metal-air batteries and fuel cells in the future, it is a great challenge to prepare these bifunctional electrocatalysts based on transition metal oxides on a large scale. W. Wang et al. reported the preparation of mesoporous $MnCo_2O_4$ spinel ORR/OER bifunctional electrocatalysts on a large scale, exhibiting ORR/OER overpotentials as low as 0.83 V due to their surface rich in Mn^{4+} and Co^{2+} [153]. It has been reported that Cu-doped Mn_2O_3 nanospheres have been synthesized on a large scale by a microwave-assisted route as a bifunctional electrocatalyst for alkaline fuel cells. Successful large-scale preparation of single-crystal CoO nanorods as efficient bifunctional electrocatalysts for all-solid-state zinc-air batteries, generating oxygen vacancies and highly active surfaces on electronically conductive substrates, resulting in low charge–discharge overpotentials and high energy density and excellent electrochemical performance for long-term stable cycling 353. Recently, amorphous NiFeMo oxides (up to 515 g per batch) were synthesized on a large scale by a simple supersaturated co-precipitation method, and the as-prepared amorphous NiFeMo oxides exhibited uniform elemental distribution and excellent OER activity, in 0.1 M KOH, the overpotential of 10 mA cm^{-2} is 280 mV.

3.3.2 Transition Metal Hydroxides

At present, the catalyst for oxygen evolution reaction with superior catalytic effect is layered double metal hydroxide (LDH). Layered double metal hydroxide is also known as hydrotalcite-like material. The main layer of LDH is positively charged, the interlayer ions are negatively charged, and the layer and interlayer ions are assembled into compounds through non-covalent interactions. Its structure and composition have the following characteristics: (1) the composition of metal ions in the main laminate is controllable; (2) the quantity and type of intercalated anions can be adjusted; (3) the particle size and distribution of the intercalated assembly can be controlled. From the characteristics and composition of this type of material, it can be found that LDH materials with various structures and compositions and different properties can be prepared by adjusting the types and proportions of laminate elements and selecting other anions for interlayer intercalation. Multiple types of LDHs provide ideas and possibilities to tune their electrocatalytic performance for oxygen reactions.

Including one or more transition metal ions in the laminate can effectively improve the electrochemical catalytic performance of LDH materials. Asefa et al. [155] synthesized a binary Zn-Co LDH by co-precipitation of sodium hydroxide alkaline solution and metal salt and applied it to the alkaline oxygen evolution reaction, and further employed hydrogen peroxide to oxidize some of the Co^{2+} in the laminate to Co^{3+}. The experimental test results showed that, compared with the traditional metal hydroxide cobalt hydroxide and metal oxide Co_3O_4, the binary Zn-Co LDH exhibited superior oxygen evolution catalytic performance with a lower overpotential of the oxygen evolution reaction and higher Turnover Frequency. Studies have shown that electrochemically inert Zn^{2+} can be regarded as a support and provide synergy in the catalytic system. H. Lin et al. also used a similar method to synthesize binary Ni-Co LDH and applied it to the oxygen evolution reaction. The Ni-Co LDH exhibited good catalytic performance for the oxygen evolution reaction. More Co^{3+} active sites and the rapid migration of ions and electrons within the layered structure jointly promoted the improvement of material properties. In addition to Co-based LDHs, LDHs containing Ni or Fe, and LDHs containing both Ni and Fe are widely considered to have high OER catalytic activity. With excellent catalytic performance, a complete research idea has been formed for Ni-Fe LDH. Specifically, it includes the preparation, catalytic characterization, structure and morphology characterization of Ni-Fe LDH, as well as the discussion and study of the reaction mechanism of oxygen evolution. The preparation methods of Ni-Fe LDH mainly include co-precipitation, solvothermal, electrodeposition, and so on.

Based on the determined LDH layer metal and its ratio, to obtain LDH with higher activity and better catalytic performance, new regulations and research on its structure and morphology have been carried out. As a typical two-dimensional material, identifying catalytically active sites of LDH is a research hotspot. Based on recent research showing that metal sites exposed at the edge of the laminate are more active than metal sites surrounded by six oxygen atoms in the closely packed laminate, how to make more exposed metal sites have attracted extensive attention of scholars. The current research is mainly to expose more active metal sites through two methods. One is to exfoliate the closely packed layered LDH into a single-layer or several-layer structure to expose more metal sites in the layered plate. The other is to introduce vacancies or defects in the layer of LDH so that the metal sites are unsaturated coordination, and these unsaturated coordination metal ions are also considered to have higher activity. Despite the high intrinsic activity of LDH, its poor electrical conductivity has hindered its efficient application in alkaline oxygen evolution reactions. Previous studies have found that electron transport is significantly enhanced at the interface with an asymmetric structure. Still, this asymmetric structure has a minimal range of action (several nanometers) and cannot achieve accelerated electron transfer throughout the material. High-performance LDH needs to be explored urgently, or the composite material of high-conductivity carrier has also become the focus of research.

3.3.3 Transition Metal Sulphur Compounds

Transition metal sulfide nanomaterials have received extensive attention due to their wide application in various energy devices including fuel cells, solar cells, sensors, lithium-ion batteries, supercapacitors, etc., attracting more and more scholars to investigate for improving their performance. As a promising non-precious metal catalyst, transition metal sulfides are expected to become new energy materials to replace Pt-based catalysts and Ru-based catalysts. Recently, layered metal sulfides have been successfully used as electrode materials for lithium-ion batteries and supercapacitors. The broad application of transition metal sulfides has benefited from advances in synthesizing nanostructured materials with different sizes and morphologies. When the size of the system is reduced to the nanometer scale, the quantum size effect will occur, leading to essential changes in its physical and chemical properties. In addition, transition metal sulfide nanomaterials can provide higher specific surface area and enhance the effective contact of the reaction interface, which is beneficial to energy conversion.

T. Huang et al. synthesized high-purity cubic NiS_2 by a simple one-pot hydrothermal method. Its catalytic ORR half-wave potential is only 40 mV lower than that of Pt/C, but the limiting current intensity is higher, and the stability is more robust [156]. L. Zhu et al. prepared $(Co, Ni)S_2$ thin films by magnetron sputtering, which had an ORR catalytic activity close to that of Pt [157]. Cobalt sulfide has good catalytic activity for OER in alkaline medium with good chemical stability and electrical conductivity. Studies have shown that incorporating Co atoms into the edge positions of other transition metal chalcogenides decreases the free energies of catalysts for adsorption of H and O-containing species. S. Shit et al. formed CoS_x/Ni_3S_2@NF composites with cobalt–nickel sulfides anchored in nickel foam, as electrocatalysts, in 0.1 M KOH electrolyte, at a current density of 20 mA cm^{-2}, the catalytic OER reaction overpotential was only 280 mV, which was stronger than that of Ni_3S_2@NF, indicating that the bimetallic synergy is beneficial to the improvement of catalyst performance [158]. Transition metal phosphides, with suitable d-electron configuration and abundant chemical states and zero-valent metal-like properties, are another attractive class of electrocatalysts. X. Guo et al. used butterfly wing scales as a template to prepare an amorphous Ni-P catalyst with a dendritic hollow structure, which catalyzed an OER reaction potential of 1.63 V at 10 mA cm^{-2} in acidic electrolyte [159].

Transition metal sulfide/carbon-based materials include many bifunctional electrocatalysts for oxygen reactions. Transition metal sulfides are a class of compounds with diverse crystal structures and complex electronic state compositions. Compared with oxides, they have lower resistance and better stability, making them widely used in the field of electrocatalysts. C. Liu et al. prepared a composite material composed of highly dispersed CoS_x nanocrystals and N-doped porous carbon (CoS_x@NMC) [160]. The synergistic effect of this composite shows that the material has excellent ORR/OER dual catalytic performance. The open-circuit voltage of the zinc-air battery prepared based on the catalyst is 1.44 V, the peak power density is as high as 269.7 mW cm^{-2}, and the constant current (2 mA cm^{-2}) charging-discharging cycles up to 1288 times. K. Wan et al. reported a mesoporous nickel sulfide NiS_x (NiS and Ni_3S_4) supported on N-doped mesoporous carbon composites (NiS_x/NMC), which catalyzed the ORR reaction with a half-wave potential as high as 0.89 V [160]. As a bifunctional catalyst, the catalytic OER reaction potential is only 1.50 V (10 mA cm^{-2}). The peak power density and energy density of the prepared liquid zinc-air battery are as high as 186 mW cm^{-2} and 805 Wh kg^{-1}, respectively, and can be stably charged and discharged for 300 cycles.

There are currently many methods for preparing transition metal sulfide nanomaterials, among which the main preparation methods include chemical exfoliation, single-source precursor, electrodeposition, and hydrothermal/solvothermal method. The so-called chemical peeling method refers to a technique in which a swelling agent is inserted between the layers of the object to be peeled by a specific chemical method. Then the peeled material is peeled off in a certain way. Generally, to obtain layered nanomaterials, a chemical exfoliation method is used to prepare monolayer compounds, thereby realizing the preparation of monolayer structures and the preparation of nanosheets. Chemical peeling is very effective for materials that are difficult to peel physically, but it is not widely used due to its strict selection of expansion agents.

The single-source precursor method is a preparation method in which the selected precursor is processed to obtain the corresponding transition metal sulfide material. Usually, the chosen precursor must contain the desired metal and the corresponding sulfur element so that the related compound can be directly decomposed at a specific temperature. By adjusting the ratio of the original reagents (including precursors, ligands and solvents, etc.), as well as the reaction temperature and reaction time, the morphology and size of the synthesized materials can be precisely controlled. Up to now, many transition metal sulfides are prepared by this method. Electrodeposition refers to the process in which metals, alloys or metal compounds are deposited on the electrode surface from the solution of their compounds under the action of an electric field to form a catalyst layer, usually accompanied by the gain and loss of electrons. There are many methods for preparing transition metal sulfides by electrochemical deposition, including cyclic voltammetry, pulse electrodeposition, square wave scanning method, potentiostatic/galvanostatic method and so on. The advantages of electrodeposition preparation technology lie in mild conditions, simple operation, and easy control. By simply controlling the parameters of electrodeposition, such as deposition voltage, reaction time, and choice of substrate, the nucleation rate of the material can be controlled to further regulate the morphology and structure of the product and the thickness of the deposited film. Many sulfide nanostructures or thin films are prepared by electrodeposition and hydrothermal/solvothermal methods. In recent years, the hydrothermal/solvothermal method has been the main method for synthesizing transition metal sulfide nanomaterials. Under the action of high pressure, the reactants in the reaction vessel are dissolved, recrystallized and grown to obtain the desired reaction product. Due to its unique homogeneous nucleation and heterogeneous nucleation mechanisms, the hydrothermal/solvothermal method produces electrocatalysts with high purity, easy particle size control, and good dispersibility. So far, hydrothermal/solvothermal methods have made significant progress in synthesizing transition metal sulfides with various nanostructures, such as MoS_2 hollow cubic cage structure, CuS microtubule structure, FeS_2 nanomesh, CoS flower structure, etc. There are many factors that affect the hydrothermal reaction and solvothermal reaction, such as solution pH value, reactant concentration, reaction time and so on.

Transition metal chalcogenides are constructed by the interaction of metals (e.g., Fe, Co, Ni) and chalcogen atoms (e.g., S, Se) to form unique crystal structures that are beneficial for electron transfer and oxygen adsorption in oxygen-catalyzed reactions. Recent studies have shown that MOFs can be considered promising precursors for synthesizing porous carbon-based metal sulfide bifunctional catalysts. For example, X. Wang et al. synthesized trimetallic CoNiFe sulfide mesoporous nanospheres (CoNiFe-SMNs) with uniform morphology using a trimetallic MOF as a precursor and a subsequent thioacetamide anion exchange method (Fig. 3.10a) [161]. CoNiFe-SMNs exhibited a low η_{10} of 1.429 V in terms of OER and a low ΔE of 0.71 V between η_{10} and $E_{1/2}$ due to the synergistic effect of the trimetal ions. The rechargeable ZAB based on CoNi-SMNs also exhibited remarkable cycling stability (low voltage gap of 0.76 V at 2 mA cm^{-2}) (Fig. 3.10b) and high power density of 140 mW cm^{-2}. This approach could pave the way for designing other MOF-derived metal sulfides with better catalytic activity.

Besides sulfurization, selenization can also be used to convert these MOF precursors into carbon-based metal selenides. For example, M. Cao et al. demonstrated the in-situ coupling of $Co_{0.85}Se$ nanocrystals with N-doped carbon (denoted as $Co_{0.85}$-Se@NC) by direct selenization of ZIF-67 polyhedra (Fig. 3.10c) [162]. Owing to the synergistic effect between $Co_{0.85}Se$ nanocrystals, N-doped carbon and Co-N-C structure, the obtained $Co_{0.85}Se@NC$ catalyst exhibited excellent bifunctional catalytic performance with a small η_{10} of 1.46 V for OER and ORR A positive $E_{1/2}$ of 0.814 V (Fig. 3.10d). Furthermore, the obtained ZABs with $Co_{0.85}Se@NC$ as the air cathode exhibited a high peak power density of 268 mW cm^{-2} (Fig. 3.10e), a low voltage gap of 0.80 V, and long cycle life of over 180 times. This work will provide a facile and feasible strategy for the selenization of MOF materials with significantly improved catalytic activity and long-term cycling stability.

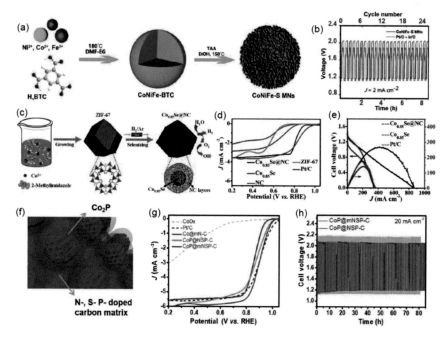

Fig. 3.10 **a** Schematic diagram of the synthetic route of CoNiFe-SMNs. **b** Electrodynamic charge–discharge and power density curves of CoNiFe-SMNs and Pt/C + Ir/C [161]. **c** Illustration of the fabrication process of $Co_{0.85}$-Se@NC. **d** Spinning disk electrode LSVs. **e** Polarization curve and corresponding power density plot of the battery [162]. **f** 1D cobalt phosphide intercalated into nitrogen, sulfur, and phosphorus co-doped carbon matrix. **g** Polarization curves of various cobalt-based hybrid catalysts and Pt/C. **h** CoP@NSP-C and CoP@mNSP-C air electrodes at 20 mA cm^{-2} [163]

The combination of transition metal phosphides with conductive carbon matrix can be obtained through phosphating process using MOFs as templates. For example, A. Manthiram et al. prepared one-dimensional mixed catalysts (CoP@mNSP-C) consisting of ultrasmall cobalt phosphide nanoparticles embedded in N-, S-, and P-doped carbon supports by combining a MOF-derived transformation approach with an in situ cross-linking approach (Fig. 3.10f) [163]. During the in situ synthesis, the one-dimensional cobalt precursor reacted with 2-methylimidazole to form ultra-small ZIF-67 nanocrystals, and the introduced triethylamine reacted with poly(cyclotriphospholipid-co-4,40-sulfonyldicarbonate) phenol) (PZS) cross-linking. Owing to the high content of CoP and very small nanoparticle size in CoP@mNSP-C, the 1D hybrid catalyst exhibited remarkable bifunctional catalytic activity with a positive $E_{1/2}$ of ORR of 0.90 V (Fig. 3.10g) and a low OER η_{10} is 1.64 V. Furthermore, the rechargeable ZABs based on CoP@mNSP-C exhibited a low potential gap of 0.86 V and excellent cycling stability with negligible voltage change over 100 cycles over 80 h (Fig. 3.10h). This new approach will provide an indicator for designing metal phosphide catalysts with remarkable bifunctional catalytic activity and excellent cycling stability.

In conclusion, appropriate metal ions and organic ligands can coordinate with each other to construct crystalline porous MOFs with ordered crystalline porous structures, abundant available atomically dispersed metal active sites, and high surface areas. Although these advantages endow pristine MOFs with high catalytic activity, selectivity, and fast mass transfer, their structural instability and low electrical conductivity in harsh acid/base environments and the use of high-cost conducting ligands largely limit their practical application in ZABs. To overcome this problem, controllable chemical/thermal treatment of MOFs has been employed to prepare porous metal-doped carbon materials with tunable nanostructures and chemical compositions. These MOF-derived materials combine the advantageous features of MOFs (e.g., porosity/high surface area) and metal-doped carbons (e.g., abundant active metal sites/ideal conductivity) with high catalytic activity, perfect selectivity, good electrical conductivity and excellent structural stability.

3.3.4 Transition Metal Single-Atom Catalysts

Introducing metal elements into carbon-based nanomaterials is another effective way to improve electrocatalytic activity. Carbon nanomaterials can promote the dispersion and utilization of effective catalytic sites, thereby enhancing catalytic activity. Heteroatom-doped carbon materials possess high specific surface area, high electrical conductivity, and sufficient functional groups to provide intimate interactions with metal-based materials. This interaction can effectively suppress the cathodic corrosion process, resulting in high durability and chemical stability of the catalyst. TM (e.g., Fe, Co, Ni, Cu, Mo, and W) and nitrogen-co-doped carbon-based (M-N-C)-based or carbon-containing catalysts have been shown to be one of the most promising bifunctional candidates for OER and ORR. Intrinsic carbon defect sites, such as vacancies or topological defects, can also provide unique sites for trapping metal species. The large difference in electronegativity between TMs and carbon atoms produces sufficient charge transfer to make these building blocks (isolated metal species trapped by intrinsic defects) electrocatalytic active centers. Different activated carbon sites have inherent defects or heteroatom doping. Incorporating these metal species is generally considered to change the electronic structure and catalytic properties of adjacent carbon atoms and act as active catalytic centers or participate in construction. Single-atom catalysts (SACs) are highly dispersed metal atoms immobilized on supports, which can maximize the utilization efficiency of metal atoms, thereby achieving high activity, stability, and selectivity of electrocatalytic reactions. SACs with $M-N_x-C$ sites on the carbon matrix of ORR and OER have been extensively studied due to their optimal atom utilization efficiency and high intrinsic activity. Another method for introducing metallic elements into carbon materials is hybrid electrocatalysts, which consist of transition metal-based particles (including metals, alloys, oxides, nitrides, sulfides, etc.) and carbon supports.

In addition, metal–organic frameworks (MOFs) and covalent organic frameworks (COFs) have also been considered as ideal precursors for the rational dispersion of metals in carbon scaffolds.

Although transition metal doped carbon matrix has good catalytic activity for ORR, studies have shown that the metal atoms inside the nanoparticles cannot be effectively utilized. By reducing the size of the active metal nanoparticles, the metal utilization efficiency can be improved, leading to high activity and durability. Furthermore, tuning the coordination chemical environment of the central metal is simpler for monodisperse metal atoms. Therefore, single-atom catalysts (SACs) have been developed as emerging catalysts. The metals in SACs are dispersed on the substrate in the state of single atoms, which have the advantages of unique electronic effects, uniform distribution of active centers, and maximum utilization efficiency of metals. Since MOFs can generate uniform M-NQ active centers and porous structures after pyrolysis, MOFs are commonly used as precursors for the preparation of SACs. Different templates (SiO_2 spheres, FeOOH nanorods, MgO, etc.) can also be used to prepare SACs.

Iron-based and cobalt-based M-N-C materials are the most representative single-atom ORR and OER electrocatalysts. Y. Wang et al. developed Fe/N co-doped multi-layer porous carbon (Fe/N/C) through carbonization and etching steps. Fe/N/C exhibited excellent ORR activity in alkaline media ($E_{1/2} = 0.902$ V vs. RHE) and OER performance ($E_{j10} = 1.66$ V) [164]. Z. Li et al. synthesized a single-atom Fe-N_x-C electrocatalyst by in-situ incorporating Fe-Phen complexes into nanocages during ZIF-8 growth (Fig. 3.11a) [165]. The results show that pyridine nitrogen, Fe-N_x and graphitic nitrogen coexist in the Fe-N_x-C catalyst (Fig. 3.11b). C and N provide coordination to Fe atoms in the form of Fe-N_x (Fig. 3.11c). The graphitic nitrogen affects the geometrical and electronic structure of the carbon framework and improves the limiting current density of the catalyst for ORR. The ΔE ($\Delta E = E_{j10} - E_{1/2}$) of Fe-$N_x$-C is 0.92 V, which is better than the calculated value of Pt/C for ORR ($E_{1/2}$) and RuO_2 for OER (E_{j10}) ($\Delta E = 0.94$ V), indicating that Fe-N_x-C has better bifunctional activity (Fig. 3.11d). Under the Fe-N_x-C catalyst, ZAB was applied in ZAB at a current density of 10 mA cm^{-2} for more than 250 h (Fig. 3.11e), indicating that single-atom Fe-N_x-C electrocatalyst is a promising rechargeable ZABs. Co atoms with similar properties are also trapped by inherent defects. Z. Liu et al. proposed a method for preparing highly dispersed Co-N-C catalysts by anchoring metal atoms in N-doped carbon via a defect strategy and following carbon coating [166]. The results show that the unique structure promotes the increase of defects in carbon, enhances the dispersion of Co, and increases the number of active sites. The synergistic effect of Co-N, pyridine N, and graphitic N endows it with excellent catalytic performance. S. Mu et al. synthesized an advanced CoN$_x$-C carbon framework from high-surface-area 3D ZIF nanocrystals and performed ORR and OER tests on it [167]. Due to the synergistic effect of abundant Co-N coupling centers, the nanorods possess a unique high specific surface area structure, abundant porosity and abundant active sites, exhibiting remarkable bifunctional ORR/OER activity ($E_{j=10}$(OER)-$E_{1/2}$ (ORR) ≈ 0.65 V). The resulting TM-N-C active sites can significantly enhance the ORR/OER catalytic activity of doped carbon materials. In summary, the bimetallic

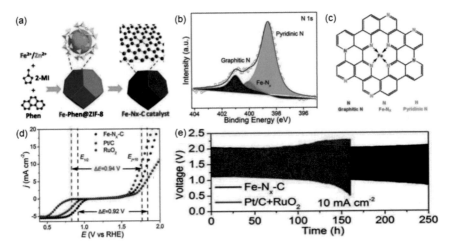

Fig. 3.11 a Schematic diagram of the synthesis process of noble metal-free single-atom Fe-N$_x$-C electrocatalyst. **b** XPS spectrum of N 1 s. **c** Illustrates the N species in the carbon framework. **d** Overall polarization curves of Fe-N$_x$-C, commercial Pt/C, and commercial RuO$_2$ in 0.1 M KOH. **e** Cycling tests were performed with Fe-N$_x$-C and Pt/C + RuO$_2$ mixture catalysts at a current density of 10 mA cm^{-2}, respectively [165]

co-doped carbon material is considered to be a very bright bifunctional catalyst with abundant active sites. H. Liu et al. developed a novel and versatile ionothermal synthesis method to fabricate 2D CoFe/SN-doped, tailored open-pore porous carbon nanosheets [168]. It is found that the synergistic contribution of Fe-N$_4$ and Co-N$_4$ active sites greatly enhances the bifunctional catalytic performance due to the slightly weakened adsorption strength of the ORR/OER intermediate on the active sites.

Metal nitrogen-doped carbon materials (M-N-C, M=Fe, Co, Ni, etc.), consisting of M-N$_x$ species embedded in a carbon matrix, have been widely reported due to their superior ORR catalytic activity. DFT calculations indicated that the M-N$_4$ site could be as active as Pt for O adsorption and O–O bond disruption during the ORR process. Furthermore, the MOF-derived M-N-C materials composed of carbonized ligands with abundant N and transition metal sites exhibit outstanding ORR activity and excellent OER properties, thus attracting extensive attention in the field of catalysis.

Single atoms that reduce the size of metal particles to the atomic scale and embed them in ligand-derived porous carbon matrices are critical to achieving abundant active sites. Due to neat dispersion and high atom utilization efficiency, MOF-derived single-atom M-N-C catalysts generally exhibit remarkable catalytic activity for ORR and OER. In recent studies, MOF-derived N-doped carbon substrates were selected as carbon supports for Co single atoms to develop active Co-N$_4$ sites for oxygen catalysis. For example, L. Huang et al. reported a salt-templated approach to preparing highly dense and monodisperse Co monolayers on N-doped 2D carbon nanosheets (denoted as SCoNCs) by direct pyrolysis of ZIF-67-encapsulated KCl particles atom (Fig. 3.12a) [169]. Benefiting from the unique structure and the highly dispersed

single Co-N species in the carbon matrix, the resulting SCoNCs showed highly efficient bifunctional catalytic activity with an $E_{1/2}$ of 0.91 V for ORR (Fig. 3.12b) and
a low E_{10} of 1.54 V for OER. The ZABs based on the SCoNC catalyst also exhibited a high peak power density of 194 mW cm^{-2} (Fig. 3.12c) and a high energy
density of 945 Wh kg$_{Zn}^{-1}$. In addition, S. Guo et al. also successfully synthesized a
bifunctional single-atom catalyst (Co SA@NCF/CNF), which consisted of Co single
atoms intercalated into N through a general impregnation-carbonization-acidification
route [170]. In the doped porous carbon sheets, ZIF and electrospun nanofibers are
used as precursors. Owing to the unique hierarchical porous structure and abundant
single-atom active sites, Co SA@NCF/CNF exhibited excellent bifunctional catalytic
activity with ΔE values as low as 0.75 V (Fig. 3.12d). The wearable ZAB device based
on Co SA@NCF/CNF exhibited good stability under deformation (Fig. 3.12e) and a
satisfactory battery capacity of 530.17 mAh g$_{Zn}^{-1}$ (Fig. 3.12f). Therefore, ZIF-67 can
be reduced in situ during pyrolysis to form atomically dispersed Co-N$_x$ sites with
good bifunctional ORR/OER catalytic performance, good electrical conductivity,
and excellent stability.

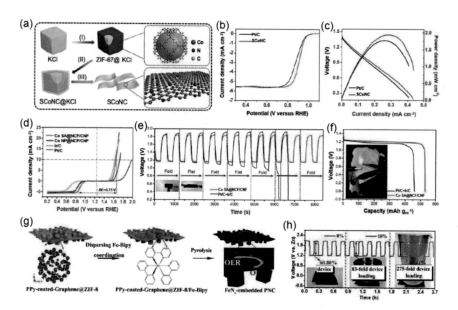

Fig. 3.12 **a** Schematic diagram of the synthesis procedure of SCoNC catalyst. **b** LSV measurement
of the as-prepared catalyst and 20% Pt/C at 1600 rpm. **c** Polarization and power density curves of
primary ZABs using Pt/C and SCoNCs as ORR catalysts [169]. **d** ORR/OER bifunctional LSV
curve. **e** Charge–discharge curves of the wearable ZAB under alternating folding and releasing
conditions. **f** Voltage-capacity curves of the as-prepared ZABs. Inset: A digital photograph showing
a practical application of a wearable ZAB [170]. **g** Schematic diagram of the synthesis of FeN$_x$
embedded in 2D porous nitrogen-doped carbon. **h** Electrostatic discharge–charge cycling curves at
5 mA cm^{-2}, showing the behavior at different compressive strains [171]

Besides Co-N-C materials, single-atom Fe-N-C catalysts are also representative bifunctional catalysts. Fe-N_x species are considered as the main active components for oxygen-catalyzed reactions in alkaline media. For example, L. Ma et al. synthesized a bifunctional catalyst (denoted as FeN$_x$-PNC) by pyrolyzing a mixture of PPy-coated graphene and ZIF-8/Fe-Bipy, [171] which combined single-site active Fe-N_x species are intercalated into highly graphitized two-dimensional porous N-doped carbon (PNC) (Fig. 3.12g). Furthermore, due to the high specific surface area, good electrical conductivity, and abundant single-site active Fe-N_x species of 2D porous PNCs, the resulting FeN$_x$-embedded PNCs exhibited excellent bifunctional catalytic activity, with $E_{1/2}$ is 0.86 V and E_{10} for OER was 1.62 V. Furthermore, when assembled into a solid-state rechargeable ZAB, it exhibited excellent battery performance with a voltage gap as low as 0.78 V. Furthermore, it can reach a compressive strain of 54% and bend to 90° without significant voltage decay (Fig. 3.12h). Therefore, ZIF-8 is one of the most efficient candidate templates to form hierarchical porous carbon substrates supporting metal single atoms to develop M-N_4 sites with high catalytic activity for oxygen catalysis.

Among the bulk MOF-derived M-N-C materials, Co-N-C materials, i.e., Co nanoparticles embedded in the N-doped carbon matrix, generally exhibit excellent electrocatalytic activity for ORR and OER. For example, M. Zhang et al. rationally designed and reported a novel Co@N-C bifunctional catalyst (C-MOF-C2-T) by pyrolyzing a pair of enantiomeric chiral 3D MOFs, as shown in Fig. 3.13a [172]. C-MOFC2-900 has a unique hierarchical rod-like structure composed of Co nanoparticles uniformly intercalated into N-doped carbon rings. Due to its unique design and multiple active sites containing metallic Co, CoN_x, CoO_x and N species, the results of C-MOFC-C2-900 for both ORR ($E_{1/2} = 0.817$ V) and OER ($E_{10} = 1.58$ V) showed outstanding electrocatalytic activity. When assembled into ZAB, it exhibited excellent battery performance with a low potential gap of 0.53 V and cycling stability without obvious deterioration (Fig. 3.13b). Such a small change makes it superior to commercial materials and one of the most durable materials in ZAB.

In addition, researchers are also working on designing other types of MOF precursors to construct Co-N-C materials as bifunctional catalysts. For example, M. Wu et al. synthesized neodymium-doped carbon/cobalt nanoparticles/neodymium-doped carbon (denoted as NdC-CoNPs-NdC) multilayer interlayer nanobrids by pyrolyzing a novel Co-MOF with a three-dimensional molecular structure (Fig. 3.13c) [173]. Due to the synergistic effect of monodisperse Co nanoparticles with high activity/stability and porous N-self-doped carbon framework with high electrical conductivity, the obtained NdC-CoNPs-NdC nanocomposites exhibited excellent bifunctional catalytic activity. When the digital carbon cloth coated with NdC-CoNPs-NdC as air cathode, Zn foil as anode, and polyacrylamide-polyacrylic acid/6 M KOH alkaline gel as solid electrolyte were assembled into on-chip all-solid-state rechargeable ZAB (OAR-ZAB), OAR-ZAB exhibited good cycling stability up to 150 cycles/50 h and had a specification capacity as high as 771 mAh g^{-1}. The design of this multilayer sandwich nanostructure enables transition metal/N-doped carbon nanobridges with high bifunctional catalytic activity and ZAB stability.

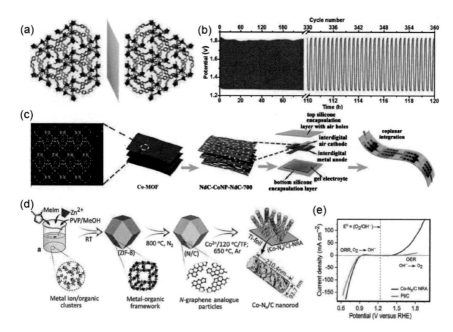

Fig. 3.13 **a** Enantiomeric packing diagram showing the three-dimensional structures of the 1L (left) and 1R (right) complexes. **b** Discharge/charge cycling curves of ZABs at 2 mA cm^{-2} using C-MOF-C2-900 as a bifunctional air electrode [172]. **c** Schematic illustration of the synthesis process of N-doped carbon/cobalt-nanoparticles/N-doped carbon multilayer interlayer nanobrids as air cathode catalysts for on-chip rechargeable ZABs [173]. **d** Synthetic route for the structural transformation of 3D ZIF nanocrystals to Co-functionalized carbon nanorod arrays. **e** ORR/OER polarization curves of Co-Nx/C NRA and Pt/C [174]

Due to their regular dodecahedral morphology and facile synthetic method, ZIFs are often used as sacrificial templates for the synthesis of Co-N-C materials. For example, S. Mu et al. fabricated advanced Co-N$_x$/C nanorod arrays (Co-N$_x$/C NRAs) by two-step pyrolysis of three-dimensional ZIF nanocrystals (Fig. 3.13d) [174]. Due to the synergistic effect of abundant Co-N coupling centers, unique nanorod structure, and high specific surface area, the resulting Co-N$_x$/C NRA catalyst exhibited more robust catalytic activity with ΔE of 0.65 V (Fig. 3.13e). Notably, when assembled into ZAB, it further showed a high energy density of 853.12 Wh kg$_{Zn}$$^{-1}$ and excellent cycling stability, which can last for 80 h at a high current density of 50 mA cm^{-2} without any obvious attenuation. Low ΔE and high energy density endow Co-N-C materials with great promise for practical application in rechargeable ZABs.

Besides 3D ZIFs, 2D ZIF nanosheets can also be used as sacrificial templates to construct Co-N-C materials. For example, Y. Guan et al. synthesized CoN-PHCNTs by pyrolysis of a novel two-dimensional bimetallic (zinc/cobalt) ZIF (two-dimensional ZIF-PHS) [175]. Benefiting from the above advantages, the Co-N-PHCNTs catalyst exhibited excellent electrocatalytic performance with a positive E$_{1/2}$ of 0.89 V for ORR and a low E$_{10}$ of 1.62 V for OER. When assembled into

ZAB, it showed good cycling stability with no obvious decay after cycling at 5 mA cm^{-2} for 673 h. This ingenious design of 2D ZIF nanosheets by a surfactant-assisted approach opens up a new direction for the preparation of 2D MOF materials.

The atomic dispersion of active sites can improve the utilization efficiency of the active sites and enhance the chemical interaction with the carriers [176]. By incorporating metal atoms into carbon materials, catalysts can simultaneously possess the intrinsic activity of metal-based active sites and the electrical conductivity of carbon. Recently, C. Zhi et al. reported Fe-N$_x$ dispersed in a carbon host, achieving excellent ORR and OER electrocatalytic reactivity [171]. The catalyst was synthesized by pyrolysis of Fe-containing ZIF-8 immobilized on polypyrrole-coated graphene. The atomic distribution of Fe-N$_x$ species was verified by high-angle annular dark-field scanning transmission electron microscopy (HAADF-STEM) images (Fig. 3.14a). The N 1 s spectrum of XPS further confirmed the existence of Fe-N$_x$ configuration (Fig. 3.14b). This unit iron-based catalyst exhibited excellent bifunctional ORR/OER activity in 0.1 M KOH solution, i.e., $\Delta E = 0.775$ V (Fig. 3.14c).

Tuning the intrinsic activity of metal-N$_x$ active sites can be achieved by controlling the atomic structure around the metal atoms. A bifunctional oxygen electrocatalyst with boron-doped Co-N-C as the active site was recently proposed [177]. Boric acid, urea, polyethylene glycol, and cobalt nitrate were mixed and pyrolyzed, followed by acid washing to remove Co nanoparticles. In the HAADF-STEM image of Fig. 3.14d, it can be seen that the atomic-scale Co species are uniformly dispersed on the carbon

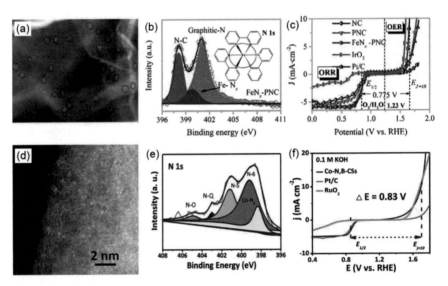

Fig. 3.14 **a** HAADF-STEM image of Fe-N$_x$ species on graphene material. **b** XPS N 1 s spectrum of Fe-N$_x$ intercalated carbon. **c** Measured at 5.0 mV s^{-1} in 0.1 M KOH solution ORR and OER LSV plots [158]. **d** HAADF-STEM image of atomic-scale Co species on a carbon sheet. **e** High-resolution XPS spectrum of N 1 s of Co/N-B doped carbon. **f** ORR and OER LSV plots measured at a scan rate of 5.0 mV s in 0.1 M KOH solution [177]

sheet. The CoN_x bond can be identified in the XPS N 1 s spectrum (Fig. 3.14e), and the B 1 s XPS spectrum confirms the formation of B-C, B-N, and B-O bonds, indicating that the active site configuration is $Co-N_x-B_y-C$. The potential gap of the boron-doped Co-N-C catalyst in 0.1 M KOH electrolyte is 0.83 V, 0.83 V for $E_{1/2}$ and 1.66 V for E_{10} (Fig. 3.14f). The electrocatalytic performance is better than that of the catalyst without boron.

Transition metal-based materials are widely used for bifunctional ORR/OER electrocatalysts. The intrinsic activity of these catalysts can be tuned by adjusting their composition and their nanostructure or by combining them with other transition metal-based materials or carbon materials. With these reasonable adjustments, excellent bifunctional activity can be achieved. The ORR and OER performance of transition metal-based M-N-C electrocatalysts are summarized in Table 3.3.

Iron-Based SACs for Zinc-Air Batteries In general, a metal-air battery consists of a metal anode (such as Li, Na, Mg, Al, K, Fe, and Zn), an air cathode, a separator, and an appropriate electrolyte solution in which zinc-air Batteries have received extensive attention due to their high energy density shown by theoretical studies [190–196]. According to the type of electrolyte, zinc-air batteries can be divided into water/liquid and solid-state types, among which zinc-air water batteries using KOH electrolytes have been known as the most mature battery systems at present [197, 198]. When

Table 3.3 ORR/OER catalytic activity of transition metal single-atom catalysts and their zinc-air battery performance

Catalysts	OER/ORR catalytic performance			Zn-air battery performance	
	Loading mass	$E_{1/2}$	η_{10}	$\Delta E@j$	Power density/cycle number (h)@j
Fe@N-C-700 [178]	0.31	0.83	0.48	0.7@10	220/100@10
C-Fe-UFR [179]	0.25	0.86	0.45	–	142/100(34)@10
N-GCNT/FeCo-3 [180]	0.20	0.92	0.5	1.49@120	89.3/240(40)@120
Cu-N@C-60 [181]	0.2986	–	–	1.19@20	210/-(100)@20
NGM-Co [182]	0.25	–	–	1.00@2	152/180(60)@2
Meso/micro-FeCo-N_x-CN-30 [183]	0.10	0.886	0.44	0.80@10	150/20(40)@10
Co-N, B-CSs [177]	0.10	0.83	–	1.35@5	100.4/128(14)@5
Fe/N/C@BMZIF [184]	1.50	0.85	0.41	0.82@10	235/100(17)@10
CoZn-NC-700 [184]	0.24	0.84	0.39	0.73@10	152/385(64)@10
Co-N-CNTs [185]	0.20	0.90	0.46	–	101/130(15)@2
3D-CNTA [186]	0.41	0.81	0.36	0.68@10	157.3/240(40)@10
3D C@Co-N-C NN-800 [187]	0.15	0.90	0.47	0.8@15	138/90(30)@15
Meso-CoNC@GF [188]	1.8	0.87	0.43	0.91@20	154.4/630(105)@20
FeN_x-embedded PNC [171]	0.14	0.86	0.39	0.78@10	278/300(55)@10
Co-DPA-C [189]	0.5	0.767	–	0.94@2	– /500(500)@2

gradually supersaturated, soluble $Zn(OH)_4{}^{2-}$ can be decomposed to produce ZnO. Meanwhile, O_2 accepts electrons released from the Zn anode, and ORR occurs at the cathode to generate OH^-. In addition, the charging process of rechargeable zinc-air batteries is also important, with oxygen evolution reaction (OER) on the air electrode and zinc electroplating on the negative electrode.

In recent years, a lot of work has been done to promote the sluggish OER/ORR process on the cathode of Zn-air batteries. Some bifunctional OER/ORR SACs have also been fabricated to assemble high-performance rechargeable Zn-air batteries [171, 199, 200]. In the synthesis of atomic $Fe-N_x/C$ catalysts for ORR, wet chemistry and subsequent pyrolysis procedures are usually performed based on defect engineering, coordination design, and steric confinement. In the synthesis of Fe SAs/MC, 5-amino-1,10-phenanthroline (phen-NH_2) ligand-coordinated Fe^{2+} was used as the Fe precursor, and the steric confinement effect of phen and SBA-15 enabled single particles Fe atoms are fixed on N defects after pyrolysis. The stable Fe SAs/MC catalyst exhibited an onset potential of 1.03 V versus RHE and a half-wave potential of 0.902 V versus RHE when directly subjected to $4e^-$ ORR in 0.1 M KOH, and the open circuit voltage of the liquid Zn-air battery assembled with Fe SAs/MC is 1.52 V versus RHE, a specific capacity of ~739 mAh g^{-1}, and an energy density of 960 Wh kg^{-1}. In the Fe SAs/MC based Zn-air batteries, there is no obvious voltage drop after three cycles, indicating the robust stability of Fe SAs/MC in Zn-air batteries. The excellent electrocatalytic performance of Fe SAs/MC for ORR and Zn-air batteries can be attributed to the strong interaction of isolated Fe with adjacent N atoms. Using SiO_2 as a hard template, $Fe-N_x$ (Fe-NCCs) embedded in carbon nanocages were synthesized after hydrothermal treatment and thermal decomposition. The coordination of Fe^{3+} with N and O forms the Fe precursor, and the carbonization eventually produces atomic $Fe-N_x$ species [201]. Before carbonization, Fe elements were distributed on the outer shell; however, during carbonization, Fe atoms were diffused, which made Fe and N elements disperse throughout the nanocages (Fig. 3.15a, b). After acid leaching, the Fe signal was drastically reduced due to the reaction of HF with Fe nanoparticles (Fig. 3.15c). In 0.1 M KOH, the high ORR activity of Fe-NCCs was demonstrated due to the formation of $Fe-N_x$ with a half-wave potential of 0.82 V versus RHE. For Fe-NCC-based Zn-air batteries, a specific capacity of 705 mAh g^{-1} and long-term stability can be achieved.

Furthermore, some atomically dispersed iron catalysts have been fabricated by pyrolysis of bimetallic zinc/iron polymers or MOFs, in which the incorporation of zinc can effectively prevent the aggregation of iron atoms. Chen and co-workers synthesized atomically dispersed $Fe-N_4$ supported on N-doped porous carbon (Fe-N_4 SAs/NPC) derived from bimetallic Zn/Fe polyphthalocyanine in rechargeable Zn-air batteries showed excellent ORR and OER performance in [185]. As shown in Fig. 3.15d, three prominent peaks appear in the N K-edge XANES spectra of $Fe-N_4$ SAs/NPC, which are attributed to the σ^* and π^* transitions of C-N molecules, and the valences of the atomic iron centers in $Fe-N_4$ SAs/NPCs range from $+2$ to $+3$ (Fig. 3.15e). EXAFS technique was used to reveal the Fe-N coordination in $Fe-N_4$ SAs/NPC, and the $Fe-N_4$ configuration was verified by EXAFS fitting (Fig. 3.15f). For alkaline ORR, the high activity (E_{onset}: 0.972 V versus RHE; $E_{1/2}$: 0.885 V

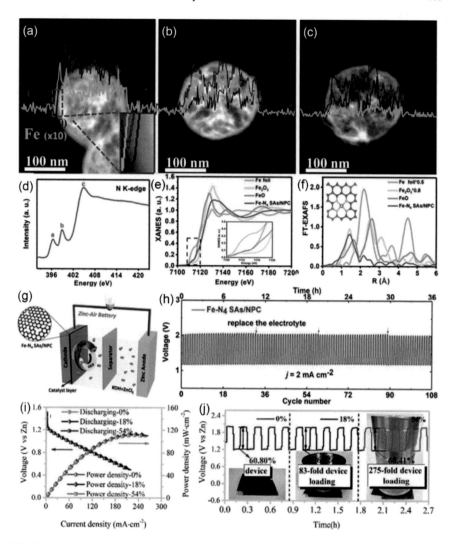

Fig. 3.15 **a–c** HAADF-STEM images and corresponding EDS line scans of hydrothermally treated products, and carbonized products before and after HF etching of Fe-NCCs [201]. **d–f** N K-edge XANES of Fe-N$_4$ SAs/NPC, Fe K-edge XANES (inset: enlarged image), and k3-weighted FT-EXAFS in R space (inset: Fe atomic structure model; C gold, N silver, Fe orange). **g** Zn-air battery device assembled using Fe-N$_4$ SAs/NPC as cathode catalyst. **h** The charge–discharge cycle performance of the Fe-N$_4$ SAs/NPC-based battery [199]. **i** Discharge polarization and corresponding power density curves of FeN$_x$-embedded PNC-based Zn-air batteries with PAM hydrogel as the electrolyte, showing 0, 18, and 54% compressive strain. **j** Cycling measurements of the rechargeability of FeN$_x$-embedded PNC-based Zn-air batteries at a current density of 5 mA cm^{-2} [171]

versus RHE) and stability of Fe-N$_4$ SAs/NPCs were demonstrated to be promising as cathode catalysts for Zn-air batteries (Fig. 3.15g). In addition, Fe-N$_4$ SAs/NPC can also be used to catalyze OER with an overpotential (η) of 430 mV and a Tafel slope of 95 mV dec^{-1} at 10 mA cm^{-2}. Using Fe-N$_4$ SAs/NPCs, a rechargeable Zn-air battery was assembled, delivering a power density of 232 mW cm^{-2} and showing negligible voltage drop over 36 h (Fig. 3.15h). Using bimetallic Fe/Co MOFs as Fe precursors, Zhi et al. prepared atomically dispersed FeN$_x$-embedded porous N-doped carbons (FeN$_x$-embedded PNCs), which exhibited high ORR activity under alkaline conditions (E$_{onset}$: 0.997 V versus RHE; E$_{1/2}$: 0.86 V versus RHE) and OER activity ($\eta = 390$ mV@10 mA cm^{-2}). Furthermore, using polyacrylamide (PAM) polymer as the electrolyte, compressible solid-state Zn-air batteries containing FeN$_x$-embedded PNCs exhibited durable stability and flexibility, as well as excellent charge–discharge capability (Fig. 3.15i, j).

For the most atomized Fe-N$_x$/C catalysts, Fe-N$_4$ sites have been shown to favorably catalyze the cathodic reaction of Zn-air batteries. Recently, Q. Xu et al. fabricated an atomically dispersed iron-based catalyst with a cantilevered structure (Fe/OES) by thermal decomposition, in which Fe-N$_4$-C molecules were regarded as active sites in the ORR process [202]. When applied in electrocatalysis, the 4e$^-$ ORR process was promoted with an onset potential of 0.997 V versus RHE and a half-wave potential of 0.85 V versus RHE. Fe/OES catalysts have been used to assemble Zn-air batteries, whose specific capacity and energy density can reach 807.5 mAh g^{-1} and 962.7 Wh kg^{-1}, respectively. For the Fe/OES-based zinc-air battery, a peak power density of 186.8 mW cm^{-2} and remarkable stability are achieved. Atomically dispersed Fe-N/C catalyst (Fe-N/C-1/30) with high ORR activity (E$_{1/2}$ = 0.895 V$_{RHE}$) was successfully prepared by Wei et al. via ball milling and subsequent pyrolysis steps and assembled Fe-N/C based zinc-air batteries [203]. It can be clearly observed from the HAADF-STEM images that the isolated iron atoms are highly dispersed in Fe-N/C-1/30 (Fig. 3.16a, b), and it can be proved from the XAS spectrum that the iron valence ranges from 0 to + 3. There are atomic Fe-N$_4$ sites (Fig. 3.16c–f). As a cathode catalyst, Fe-N/C-1/30 can help to accelerate the ORR of Zn-air batteries due to the formation of abundant Fe-N$_4$ sites, with an open circuit voltage of 1.525 V versus RHE and a maximum power density of 121.8 mW cm^{-2}. Y. Wu et al. synthesized Fe atomically dispersed on neodymium-doped carbon (Fe SAs/N-C), which exhibited high ORR activity and stability for application in zinc-air batteries [204]. It is found that the oxidation state of the isolated Fe center is 0 ~+ 2, Fe-N coordination exists in Fe SAs/N-C, and the Fe-N$_4$ configuration is proposed by EXAFS fitting. For the types of N atoms in Fe SAs/N-C, the existence of pyridinic N and graphitic N was verified by N K-edge EXAFS. Due to the formation of Fe-N$_4$ active sites, the Fe SAs/N-C electrocatalyst can help to lower the barrier energy of 4e$^-$ ORR with half-wave potentials of 0.798 V versus RHE and 0.91 V versus RHE in 0.1 M HClO$_4$ and 0.1 M KOH, respectively. As cathode catalysts, Fe SAs/N-C can accelerate ORR in Zn-air batteries with a maximum power density of 225 mW cm^{-2} and a specific capacity of about 636 mAh g^{-1}. In addition, the excellent durability of Fe SAs/N-C based Zn-air batteries is also demonstrated. By connecting in series, three Fe SAs/N-C-based Zn-air batteries can easily light up one LED panel.

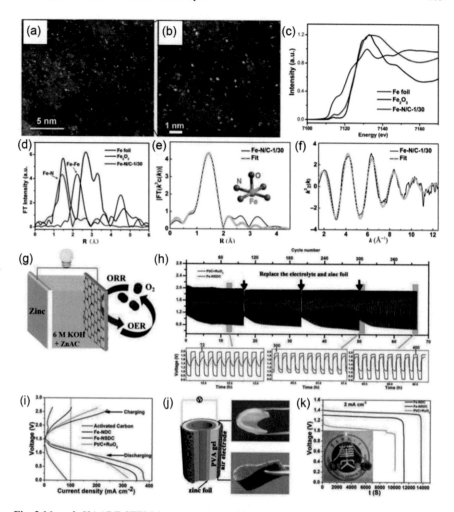

Fig. 3.16 **a**, **b** HAADF-STEM images and magnified images of Fe-N/C-1/30. **c**, **d** Fe K-edge XANES spectra and FT-EXAFS of Fe-N/C-1/30 and reference samples. **e**, **f** EXAFS fitting curves of Fe-N/C-1/30 in R space and k space [206]. **g** Schematic diagram of the working of the liquid zinc-air battery. **h**–**k** Performance of rechargeable Zn-air batteries assembled with Fe-NSDC as cathode catalyst [207]

To improve the performance of rechargeable Zn-air batteries, Z. Zhu et al. fabricated an atomical N, S-doped Fe-N-C catalyst (Fe-NSDC), in which the doped S can enhance the ORR activity by tuning the electron's structure at Fe-N$_x$ sites. Using KOH and Zn(CH$_3$COO)$_2$ as electrolytes, a Fe-NSDC-based Zn-air water battery was assembled, which exhibited good charge–discharge capability and high stability (Fig. 3.16g–i). In addition, an all-solid-state Zn-air battery can also be assembled with Fe-NSDC, in which polyvinyl alcohol (PVA) is used as the electrolyte. Solid-state batteries based on Fe-NSDC exhibit excellent flexibility and stability (Fig. 3.16j). A device consisting of two cells connected in series can light a blue light-emitting diode (Fig. 3.16k). D. Wang et al. fabricated single-atom iron (Fe-SAs/NPS-HC) supported on N, S, and P doped carbons, which can be assembled into high-performance Zn-air batteries [205]. The AC-HAADF STEM images revealed highly dispersed single iron atoms in Fe-SAs/NPS-HC and confirmed by XAS that Fe-N coordination exists in Fe-SAs/NPS-HC, where the oxidation state of the iron center is $0 \sim +3$. In addition, the Fe-N$_4$ structural model in Fe-SAs/NPS-HC was determined by DFT calculation combined with EXAFS fitting. The doped S and P heteroatoms can optimally tune the electronic structure of Fe-N$_4$ active sites, on which fast ORR kinetics can be achieved. In alkaline electrolyte, the stabilized Fe-SAs/NPS-HC catalyst exhibited excellent ORR performance with a half-wave potential of 0.912 V versus RHE, a kinetic current density of 71.9 mA cm^{-2} (at 0.85 V vs. RHE), and a Tafel slope of 36 mV dec^{-1}. Fe-SAs/NPS-HC was also successfully assembled into a Zn-air battery with Fe-SAs/NPS-HC decorated air electrode, Zn foil and 6 M KOH as cathode, anode and electrolyte, respectively. The Fe-SAs/NPS-HC based Zn-air battery exhibited an open circuit voltage of 1.45 V versus RHE, a maximum power density of 195 mW cm^{-2} and a current density of 375 mA cm^{-2}, and its long-term operational durability was also validated. Furthermore, the atomically dispersed Fe-N$_x$-C catalyst (Fe-LC-900) prepared by S. Kurungot et al. has been successfully assembled into a Zn-air battery with a maximum current density of 280 mA cm^{-2} and a peak power density of 170 mW cm^{-2}, as well as long-term stability.

In addition, bimetallic Fe-based SACs have also shown great potential in Zn-air batteries, such as atomically dispersed Fe/Co electrocatalysts with bimetallic sites [208]. Y. Wu et al. found that double Fe/Co intercalated on N-doped carbon nanotubes (Fe, Co)/CNTs was very favorable for 4e$^-$ ORR, during which the onset potential and half-wave potential in alkaline electrolyte are 1.15 V versus RHE and 0.954 V versus RHE [209]. They synthesized (Fe, Co)/CNT catalysts with highly dispersed single Fe and Co atoms (Fig. 3.17a–e). The configuration of FeCoN$_6$ in (Fe, Co)/CNT was verified by XAS analysis, which facilitated the O$_2$ activation process and enhanced the ORR performance. The stabilized (Fe, Co)/CNT catalyst can be used as an efficient cathode catalyst for Zn-air batteries with power and energy densities of 260 mW cm^{-2} and 870 Wh kg^{-1}, respectively. Similarly, A. Thomas et al. synthesized a layered meso/microporous FeCo-N$_x$/C catalyst (meso/micro FeCo-N$_x$-CN) on which the mesopores increased with the addition of SiO$_2$ templates (Fig. 3.17f, g) [183]. Due to the synergistic effect of Fe-N$_x$ and Co-N$_x$ sites, the meso/micro FeCo-N$_x$-CN exhibited excellent OER and ORR performance, which could be applied in rechargeable Zn-air batteries. When this bifunctional meso/microporous FeCo-N$_x$-CN catalyst was assembled into a liquid zinc-air battery, the open-circuit voltage reached ~1.4 V versus RHE. The maximum power density reached 150 mW cm^{-2} (Fig. 3.17h, i), and the stable cells based on meso/micro FeCo-N$_x$-CN exhibited excellent charge–discharge capability (Fig. 3.17j, k).

In conclusion, various Fe-N-C-based SACs have been widely reported, in which Fe-N$_x$ molecules, especially Fe-N$_4$ sites, are known as active sites for cathode reactions in Zn-air batteries. According to recent literature reports, massive atomically dispersed Fe-N-C catalysts exhibit excellent ORR/OER performance and hold great

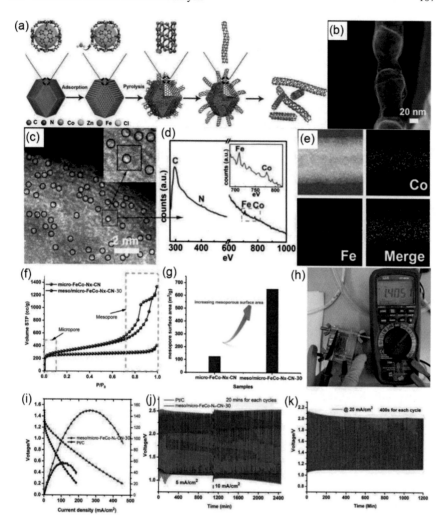

Fig. 3.17 **a** Schematic diagram of the preparation of (Fe, Co)/CNT. **b, c** HAADF-STEM and enlarged HAADF-STEM images of (Fe, Co)/CNT. **d, e** EEL spectra on the red rectangle in (**c**) and the corresponding EELS mapping images of Co, Fe, and N [209]. **f, g** Adsorption/desorption isotherms of N_2 and calculated mesoporous surface areas from the t-plots of the meso/micro FeCo-Nx-CN-30 catalysts. **h** Open circuit potential of Zn-air battery loaded with muon/micron-Nx-CN-30. **i** Discharge curves and corresponding power densities of Zn-air batteries loaded with muon/micron-FeCo-Nx-CN-30 and Pt/C catalysts. **j** Cycling performance of Zn-air batteries using meson/micron-Nx-CN-30 and Pt/C catalysts at 5 mA cm^{-2} and 10 mA cm^{-2} (20 min per cycle). **k** Cycling performance of a meson/micron-Nx-CN-30-based battery for 20 h at 20 mA cm^{-2} [183]

promise in liquid or solid Zn-air batteries. For the excellent Fe-N-C catalyzed Zn-air battery, high open circuit, peak power density and stability are achieved under practical operating conditions. However, the long-term operational durability of batteries still needs to be improved, which plays a key role in practical applications. Notably, due to different metal loadings and intrinsic activities of single-atom sites, FeN_x molecule-based SACs may exhibit quite different catalytic activities in terms of OER and ORR, which is largely related to the synthetic strategy. Furthermore, Fe-based SACs at bimetallic sites have been developed, and most of them show better ORR/OER activities compared to SACs as single metal sites. As mentioned above, the catalytic performance of atomically dispersed FeCo-based catalysts in Zn-air batteries can be significantly enhanced due to the synergistic effect of $Fe-N_x$ and $Co-N_x$ sites.

Besides $Fe-N_x/C$-based SACs, atomic $Co-N_x/C$ materials with high ORR activity and stability have also been widely studied for Zn-air batteries [169, 177, 210, 211]. Atomically dispersed Co on N-doped carbon (NC-Co SA) prepared by S. J. Pennycook et al. has been reported to be beneficial for oxygen catalysis in zinc-air batteries [210]. Under alkaline conditions, the catalysis of NC-Co SA achieved a positive half-wave potential of 0.87 V versus RHE for ORR and a low overpotential of 360 mV for OER (@10 mA cm^{-2}) due to abundant $Co-N_x$ sites. Using NC-Co SA as the cathode catalyst, a high-performance solid-state Zn-air battery was assembled, as shown in Fig. 3.18a, b, which was flexible and stable. The NC-Co SA-based cell provided a high open voltage of 1.41 V (as shown in Fig. 3.18c, the voltage of three NC-Co SA-based Zn-air cells in series is 4.232 V), and the six LEDs could be composed of two powered NC-Co SA-based battery (Fig. 3.18c). With the help of NC-Co SA, the Zn-air battery obtained excellent charge–discharge capability with a current density of 31 mA cm^{-2} and a peak power density of 20.9 mW cm^{-2}, and its long-term durability has passed the cycle stability test was verified (Fig. 3.18d–f).

Using a surfactant-assisted approach (Fig. 3.19a, b), D. Cao et al. synthesized atomically dispersed $Co-N_4$ supported on N-doped graphene (CoN_4/NG), showing excellent performance in zinc-air batteries ORR and OER performance [212]. The atomic $Co-N_4$ structure in CoN_4/NG was determined by XAS characterization, which was highly active for alkaline ORR and OER with a half-wave potential of 0.87 V versus RHE and an overpotential of 380 mV (@10 mA cm^{-2}). Liquid and all-solid-state zinc-air batteries were assembled using CoN_4/NG. In the CoN_4/NG-based liquid battery, the open-circuit voltage and the peak power density are 1.51 V and 115 mW cm^{-2}, respectively, and it exhibits durable stability. In an all-solid-state Zn-air battery containing CoN_4/NG, excellent charge–discharge capability can be achieved with a power density of 28 mW cm^{-2} and long-term durability (Fig. 3.19c–f). Through defect engineering, Zhang and colleagues fabricated a bifunctional Co/N/O tri-doped graphene catalyst (NGM-Co) in a rechargeable zinc-air battery [182]. For NGM-Co, atomically dispersed $Co-N_x$-C active sites have been identified for ORR, where the lower electron density of adjacent C atoms is favorable for the adsorption of intermediates, further helping to improve ORR kinetics. In addition, studies have shown that N and O doping dominated in NGM-Co, making it an excellent ORR performance. Notably, NGM-Co can achieve a high limiting current density of 4.75

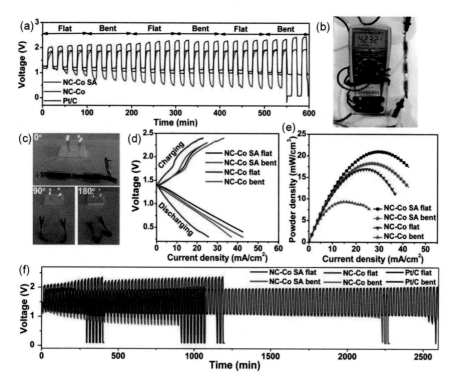

Fig. 3.18 **a** Mechanical flexibility and stability testing of NC-Co SA-based zinc-air batteries with continuous mechanical alternation between flat and curved states. **b** Three Zn-air batteries assembled using NC-Co SA as cathode catalyst with a minimum open circuit voltage of 4.232 V versus RHE. **c** Photographs of 6 LEDs powered by two NC-Co SA-based Zn-air batteries with different bending angles (0°, 90°, 180°). **d, e** Discharge and charge polarization curves, and power-current density curves of Zn-air batteries assembled with NC-Co SA and other reference samples. **f** Comparison of cycling stability of NC-Co SA based Zn-air batteries in bent and flat states [210]

mA cm^{-2} and a small Tafel slope of 58 mV dec^{-1} for ORR in alkaline. Compared with Pt/C + Ir/C, NGM-Co exhibited superior activity and rechargeability in the assembled liquid zinc-air battery, which was so stable that it could drive a toy car for a long distance. In the NGM-Co based solid-state Zn-air battery, a high open circuit voltage of 1.44 V, a high energy efficiency of 63% (at 1.0 mA cm^{-2}), and a small charge–discharge voltage gap of 0.7 V were achieved, which is also very stable and flexible.

The incorporation of B in Co-N$_x$/C (Co-N, B-CSs) can significantly reduce the energy barrier of ORR, since the addition of B can well tune the electronic structure of the active site [163]. As shown by DFT calculations, coupling the Co-N$_x$ active sites and the incorporated B atoms can facilitate O adsorption, resulting in Co-N, B-CSs with good activity in 4e$^-$ ORR (Fig. 3.19g–i). Co-N and B-CSs catalysts can be assembled into high-performance liquid zinc-air batteries with a maximum power density of 100.4 mW cm^{-2} and a voltage peak of 1.20 V at 10 mA cm^{-2}

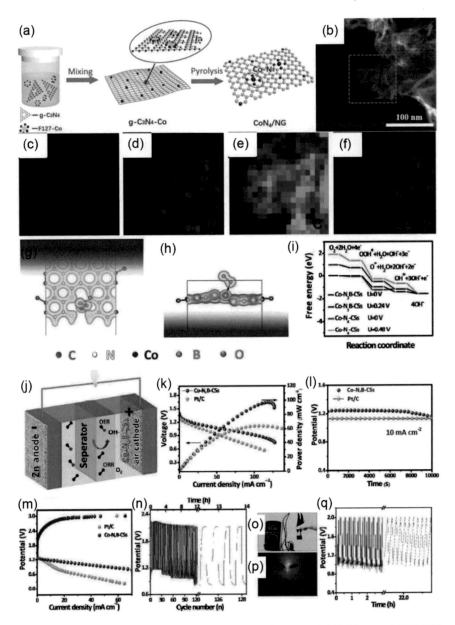

Fig. 3.19 a Schematic synthesis of single-atom CoN$_4$/NG catalyst. **b** TEM image of CoN$_4$/NG.
c–f Co, N, C and superimposed EELS maps of Co and N for CoN$_4$/NG samples [212]. **g, h** The
optimized geometry of the corresponding O adsorption configuration on the Co-N, B-CSs system. **i**
Free energy paths of the Co-N$_3$B-CSs and Co-N$_4$-CSs systems during ORR in alkaline electrolytes,
the equilibrium potentials are U = 0 V versus RHE and 0.24 V versus RHE for Co-N$_3$B-CSs; for
Co-N$_4$-CSs, U = 0 V versus RHE, 0.48 V versus RHE. **j–q** Electrochemical performance of Co-N,
B-CS based Zn-air batteries [177]

(Fig. 3.19j–n). In addition, an all-solid-state Zn-air battery with an open-circuit voltage of 1.345 V can be assembled with Co-N, B-CSs, and three series-wound solid-state batteries can power a blue LED (~3.4 V) (Fig. 3.19o, p). For Co-N, B-CS-based solid-state batteries, long-term durability was also confirmed by electrostatic cycling tests (Fig. 3.19q).

W. Hu et al. fabricated an atomically dispersed Co-Ni catalyst (CoNi-SA/NC) by pyrolysis, which exhibited high catalytic activity and stability in zinc-air batteries [202]. DFT calculations revealed that the Co-Ni double site was highly active for 4e$^-$ ORR and OER, which could facilitate the adsorption and electron transfer of the reactants in the process. When CoNi-SA/NC was used as the cathode catalyst, the durable liquid zinc-air battery provided an open circuit voltage of 1.45 V, a power density of 101.4 mW cm^{-2}, and a large specific capacity of 750.9 mAh g^{-1} and the corresponding energy density of 886.1 Wh kg^{-1}. Furthermore, the solid-state Zn-air battery assembled with CoNi-SAs/NC showed superior battery performance, and two solid-state batteries could power one LED screen. In addition, bimetallic zinc/cobalt systems have been established, which are promising in energy devices. For example, Sun et al. successfully anchored discrete Zn/Co double sites on N-doped carbon (Zn/CoN-C) for Zn-air batteries [203]. In Zn/CoN-C, the ORR activity of the Co-N$_4$ site was higher than that of the Zn-N$_4$ site, indicating that the introduction of Zn could enhance the ORR activity of the Co-N$_4$ site. Due to the high ORR activity, Zn/CoN-C can be assembled into a superior Zn-air battery with a maximum power density of 230 mW cm^{-2} and high stability. Connected in series, two Zn/CoN-C based Zn-air batteries are able to light red LEDs (~2.2 V vs. RHE) wound in series.

In most atomic Co-N-C catalysts, Co-N$_x$ sites are highly active for OER and ORR, and neighboring C atoms can tune their electronic properties. Although the ORR activity is inferior to Fe-N$_x$/C, Co-N-C based SACs show great potential in rechargeable Zn-air batteries. Furthermore, it has been proved that the incorporation of heteroatoms (such as B, P, etc.) into Co-N$_x$/C helps to tune the electronic structure of the isolated active sites, further enhancing the OER and ORR activities of Zn-air batteries. Furthermore, introducing other metals such as Ni and Zn can also improve the battery performance, which can be attributed to the synergistic effect of bimetallic sites.

So far, many other transition metal-based SACs have also been developed for zinc-air batteries. Through electrochemical measurements and theoretical calculations, atomically dispersed Ni-N-C catalysts can reduce the energy barrier of ORR, which can be regarded as promising candidates for promoting the cathode reaction of Zn-air batteries. H. Qiu et al. synthesized a bifunctional Ni-N$_x$/C-based single-atom catalyst (Ni, N codoped np-graphene) for OER and ORR that can effectively accelerate the charge and discharge of Zn-air battery cathodes procedure [213]. By electron microscopy and spectroscopic characterization, the coordination of atomically dispersed Ni with N was determined, and the formation of Ni-N bonds favored enhanced ORR kinetics (Fig. 3.20a, b). Under alkaline conditions, the Ni, N-conjugated np-graphene catalyst showed positive onset potential and half-wave potential, implying that N doping could enhance ORR performance (Fig. 3.20c). Furthermore, the synergistic effect of Ni and N not only promotes ORR but also helps to lower the energy barrier of OER with overpotentials as low as 270 mV ($@10$ mA cm^{-2}) and Tafel slopes as small as 59 mV dec^{-1}. When Ni, N co-doped np-graphene was assembled into a solid-state Zn-air battery, a maximum power density of 83.8 mW cm^{-2} was achieved (Fig. 3.20d, e), and it also confirmed the stability and flexibility of Ni, N co-doped np-graphite-based batteries (Fig. 3.20f, g).

Furthermore, some metal atoms with half- or fully-filled d-electron orbitals were selected to fabricate SACs, some of which showed excellent ORR performance in Zn-air batteries. It is known that Mn possesses a stable $3d^5 4s^2$ electron shell, which is inactive for ORR. However, when Mn is scaled down to the atomic scale, the ORR activity can be significantly enhanced, which can be attributed to the optimal state

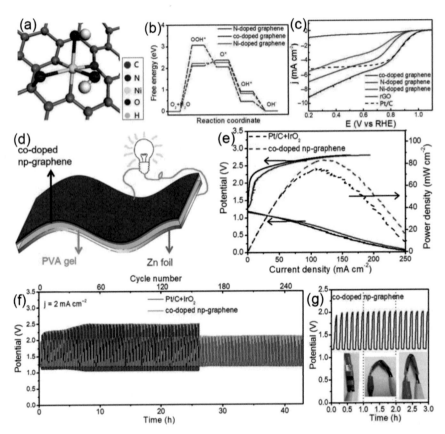

Fig. 3.20 **a** Atomic structures of O* intermediates on nickel and nitrogen doped np-graphene. **b** ORR free energy maps on nickel, nitrogen-doped np-graphene and reference samples. **c** ORR polarization curves of different electrodes in 0.1 M KOH. **d** Schematic illustration of the all-solid-state zinc-air battery doped with np-graphene. **e** Polarization and power density curves of the Zn-air battery. **f**, **g** Discharge/charge cycles of Zn-air batteries at 2 mA cm^{-2} and discharge/charge at different bending states [213]

of the $3d$ electrons of Mn. Chen and co-workers found that N, O-coordinated single-atom Mn (Mn/C-NO) can effectively catalyze ORR, in which the three-dimensional electron density of the isolated Mn center can be tuned by the coordination of N and O atoms [193]. Using DFT calculations combined with advanced characterization, it has been demonstrated that $Mn-N_3O_1$ molecules are highly active and stable towards $4e^-$ ORR, during which rational interactions between $Mn-N_3O_1$ and intermediates (including *O, *OH and *OOH species) facilitate the proton-electron transfer process. Owing to the superior ORR performance, Mn/C-NO catalysts have been successfully assembled into Zn-air batteries with current and power densities of 170 mA cm^{-2} and 120 mW cm^{-2}, respectively. In Mn/C-NO based Zn-air batteries, there was no obvious voltage drop over 20,000 s cycling, which indicated the long-term stability of Mn/C-NO in Zn-air batteries [214, 215]. Due to the inactivity of metallic Cu for ORR, Bao and co-workers designed an atomically dispersed Cu-based catalyst (Cu-NC-60) in which the three-dimensional electronic states of Cu were tuned to near the Fermi level to enhance the ORR performance [181]. By DFT calculations, the Cu valences of the proposed $Cu-N_2$ and $Cu-N_4$ configurations were determined to be $+1$ and $+2$, respectively (Fig. 3.21a), and it can be seen that $Cu-N_2$ molecules are more active than $Cu-N_4$ and $Cu-N_3$ (Fig. 3.21b). For ORR, the $Cu-N_2$ active site in Cu-NC-60 was determined from the calculated total DOS and predicted DOS (Fig. 3.21c–e). Under alkaline conditions, the Cu-NC-60 catalyst exhibited excellent ORR performance in Zn-air batteries with a current density of 142 mA cm^{-2}, a peak power density of about 210 mW cm^{-2}, and long-lasting stability.

Y. Yang et al. synthesized an atomic $Cu-N_x/C$ catalyst on which the $Cu-N_4$ sites played a dominant role for $4e^-$ ORR [214]. Under the catalysis of $Cu-N_x/C$, the stable liquid zinc-air battery exhibits a current density of 180 mA cm^{-2} and a power density of 160 mW cm^{-2}. Furthermore, atomically dispersed Zn-N-C catalysts were constructed in Zn-air batteries with significantly enhanced ORR performance compared to metallic Zn with a $3d_{10}4s_2$ electron shell [216]. By precisely controlling the heating rate, highly metal-supported Zn-N-C catalysts can be obtained for ORR, in which $Zn-N_4$ molecules are the main active sites. When the heating rate reached $1° \text{ min}^{-1}$, the Zn-N-C catalyst (Zn-N-C-1) with a metal loading of 9.33 wt% showed good performance in acidic ($E_{1/2}$, 0.746 V_{RHE}) and basic media ($E_{1/2}$, 0.873 V_{RHE}) exhibited the best ORR performance, and its activity was comparable to Fe-N-C-1 (Fig. 3.22a, b). The high stability of Zn-N-C-1 was verified by accelerated stress testing (AST) and Operando XPS, which can monitor the dynamic N configuration of Zn-N-C-1 and Fe-N-C-1 (Fig. 3.22c, d). In the XPS spectra of Zn-N-C-1 and Fe-N-C-1, there are five peaks derived from pyridine-N, $M-N_x$, pyrrole-N, graphitized N and oxide N, respectively. After 1000 cycles, the pyridine-N atoms of Zn-N-C-1 and Fe-N-C-1 were protonated in the acidic electrolyte, which can be concluded from the presence of a new peak (401.4 eV). Compared with Fe-N-C-1, Zn-N-C-1 showed a small change on pyridine-N, indicating the high stability of Zn-N-C-1 in acid. However, both Zn-NC-1 and Fe-N-C-1 are very stable under alkaline conditions. Due to the high ORR activity and stability in alkali, this optimized Zn-N-C-1 catalyst can be assembled into a Zn-air battery, showing a maximum power density of 179 mW cm^{-2} and a specific capacity of 683.3 mAh g^{-1}, the energy density

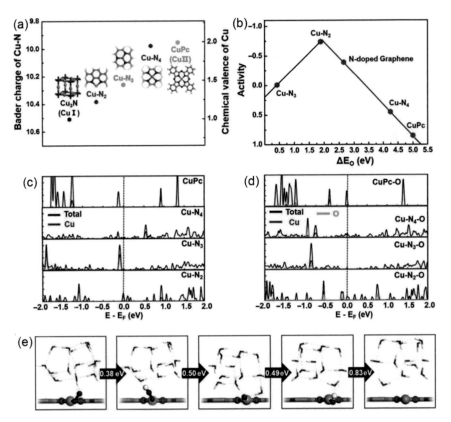

Fig. 3.21 **a** Bader charges (left order) and corresponding valences of copper atoms (right order) for different structures. **b** Volcano plot of the relationship between ORR activity and ΔE_O (binding energy of *O). **c** Total DOS (black) for different structures and projected DOS (red) of copper atoms from them. **d** The total DOS (black) of one O for different structures and the predicted DOS from their copper atoms (red) and O atoms (green). **e** ORR process on the Cu-N_2 structure [181]

is 666 Wh kg^{-1}. According to the current study, Fe-N-C and Co-N-C-based SACs exhibit superior electrocatalytic performance in Zn-air batteries compared with other carbon-based SACs. Notably, other transition metals atomically dispersed on carbon bases have been widely used to assemble Zn-air batteries, which will pave the way for the development of more efficient SACs in electrochemical conversion devices.

Besides carbon-based SACs, single atoms supported on metal oxides also show high activity and stability for ORR, which can be used to promote the cathode reaction of Zn-air batteries. For example, Y. Sun et al. fabricated atomically dispersed Ag-supported MnO_2 nanowires (Ag-MnO_2), which have been applied in rechargeable zinc-air batteries as efficient cathode catalysts [217]. Adding atomic Ag in Ag-MnO_2 can not only maximize the utilization of atoms but also induce a large number of oxygen vacancies, which is beneficial to achieving fast ORR kinetics. Due to the interaction between Ag and MnO_2, Ag doping leads to the electronic structure

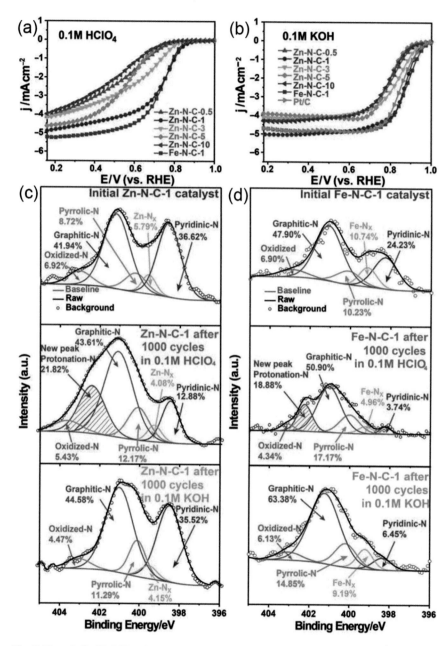

Fig. 3.22 **a, b** Zn-N-C-X and Fe-N-C-X catalysts (X = annealing rate, 0.5° min^{-1}, 1° min^{-1}, 3° min^{-1}, 5° min^{-1}, 10° min^{-1}) ORR polarization curves in 0.1 M HClO$_4$ and 0.1 M KOH, respectively. **c** High-resolution N 1 s spectra of Zn-N-C-1 before and after 1000 cycles in 0.1 M HClO$_4$ and 0.1 M KOH. **d** High-resolution N 1 s spectra of Fe-N-C-1 before and after 1000 cycles in 0.1 M HClO$_4$ and 0.1 M KOH [216]

rearrangement of the active site, which enhances the ORR activity of Ag-MnO$_2$ in Zn-air batteries. In addition, Y. Sun et al. achieved single-atom Pd anchoring on MnO$_2$ nanowires and carbon nanotube composites (Pd/MnO$_2$-CNT), which can be used as a highly efficient Zn-air battery cathode catalyst due to the synergistic effect of Pd, MnO$_2$, and CNTs [218]. Catalyzed by Pd/MnO$_2$-CNT, the stable liquid zinc-air battery exhibits excellent charge–discharge capability with a current density of 190.7 mA cm^{-2} and a peak power of 297.7 mW cm^{-2}. Using PVA as the solid electrolyte, an all-solid-state Pd/manganese dioxide-CNT battery was assembled, which can illuminate a tiny light bulb. The solid-state Zn-air batteries containing Pd/MnO$_2$-CNTs are flexible and stable, and one LED device (about 5 V) can be powered by four series-wound Pd/MnO$_2$-CNT-based batteries.

To achieve high ORR performance in Zn-air batteries, iron-based SACs have also attracted a lot of research efforts. Y. Pan et al. [199] developed single iron atoms (with a metal loading of 1.96 wt%) on N-doped porous carbon (NPC) via an in-situ anchoring strategy. The valence of each iron atom is between $+ 2$ and $+ 3$, and it cooperates with 4 N atoms to form the FeN$_4$ active site, showing the activation ability of O$_2$ (Fig. 3.23a). The as-prepared FeN$_4$ SAs/NPC showed an onset potential of 0.972 V and an E$_{1/2}$ of 0.885 V in 0.1 M KOH. The corresponding zinc-air battery showed a power density of 232 mW cm^{-2} (Fig. 3.23b) with no apparent voltage change after 108 cycles, superior to the Pt/C + Ir/C-based battery. To improve power density, Fe SAs (FeN$_4$) with metal content up to 3.8 wt% were developed on crumpled N-doped carbon nanosheets (Fe SAs/NC), yielding E$_{1/2}$ of 0.91 V in 0.1 M KOH (Fig. 3.23c) [204]. DFT calculations show that the limiting potential of Fe SAs/NC reaches 0.76 V, which is close to that of Pt (0.79 V).

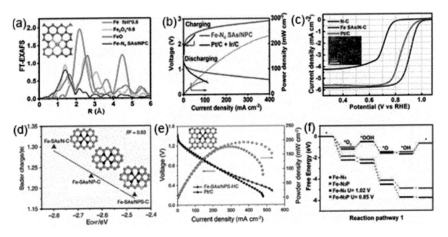

Fig. 3.23 **a** EXAFS results. **b** Charge–discharge polarization curves and power density plots [199]. **c** LSV curve. Inset: HAADF-STEM image of Fe SAs/N-C [204]. **d** Linear relationship between OH* binding energy and Bader charge. **e** Discharge polarization curve and power density plot. Inset: Schematic model of Fe-SAs/NPS-HC [205]. **f** Gibbs free energy curve [220]

Heteroatom doping in SACs has been widely adopted to improve the conductivity of carbon matrix and to change the coordination environment of the central atom [219]. As reported, the conjugated N and S atoms can induce charge redistribution in the carbon framework and enhance the bifunctional ORR/OER activity of atomically dispersed $Fe-N_x$ sites, and also $E_{1/2}$ values of 0.912 and 0.791 V were obtained in 0.1 M KOH and 0.5 M H_2SO_4, respectively. As the main active site, the isolated $Fe-N_4$ species activated and reduced O_2. The uniformly dispersed P and S atoms do not coordinate directly with the iron atoms, but modulate the electrical state through long-range interactions, which weaken the binding to the OH* intermediate to release OH^- (Fig. 3.23d). The Zn-air battery showed an OCV of 1.45 V and a power density of 195.0 mW cm^{-2} with a current density as high as 375 mA cm^{-2} (Fig. 3.23e) [205]. In addition, the $Fe-N_3P$ configuration with N, P co-ordination shows a lower original step free energy than the conventional $Fe-N_4$ site, which is more favorable for the adsorption of O_2 and the rate-determining step (OOH^- to OH^-) occurrence (Fig. 3.23f) [203]. P. Peng et al. 232 [152] coupled the graphene matrix with covalent organic frameworks (COFs) containing $Fe-N_4$ active species through intermolecular interactions. Besides the N atoms in the COFs, the Fe SAs simultaneously coordinate with the C atoms of graphene. The assembled Zn-air battery exhibits an OCV of 1.41 V, a power density of 123.43 mW cm^{-2}, long-term stability of over 300 h, and a voltage decay of less than 0.1%. Besides, oxygen dangling bonds on the surface of graphene oxide (GO) can trap metal (Fe, Co, Ni, Cu) atoms and construct corresponding $M-O_4$ sites at room temperature [153]. The mild and economic-friendly synthesis strategy operating at low temperature, which effectively reduces energy consumption, deserves our attention.

Co-N-C sites with good bifunctional ORR/OER performance have been widely used in rechargeable Zn-air batteries. B. Li et al. [154] investigated Co-POC SACs using condensed porphyrin and graphene as pyrolysis precursors. In 0.1 M KOH, Co-POC showed an $E_{1/2}$ of 0.83 V for ORR and an E_{10} of 470 mV for OER. The low ΔE of 0.87 V also demonstrated the performance of the bifunctional ORR/OER. They further designed $Co-N_x-C$ active molecules with Co loading of 1.23 wt% for flexible zinc-air batteries (Fig. 3.24a) [182]. Applying alkaline poly(vinyl alcohol) (PVA) gel as the electrolyte, the Zn-air batteries showed a high OCV of 1.439 V. They also survived long-term charge/discharge cycling and bending at different angles without significant performance changes. To improve the flexibility and wear resistance of Zn-air batteries, Guo et al. reported bifunctional $Co-N_4$ molecules (2.05 wt%) on electrospun fibers (Co SA@NCF/CNF) as binder-free air electrode [170]. $E_{1/2}$ of 0.88 V and 400 mV at 10 mA cm^{-2} were achieved in 1 M KOH (Fig. 3.24b). The wearable zinc-air battery shows an OCV of 1.41 V and a capacity of 530.17 mAh g_{Zn}^{-1}. Compared with comparable cells based on Pt/C + Ir/C, they showed a higher discharge voltage and a more stable discharge plateau (Fig. 3.24c). After frequent folding, the battery still worked well, showing resistance to external pressure (Fig. 3.24d). The rough surfaces, abundant channels, and numerous pores facilitate mass transfer at electrochemical interfaces [163]. The constructed $Co-N_3C_1$-based Zn-air battery achieved a peak power density of 255 mW cm^{-2}. The importance of combining electronic state regulation and morphology optimization is emphasized.

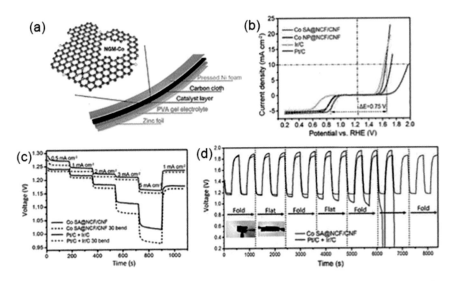

Fig. 3.24 **a** Schematic diagram of a flexible zinc-air battery [182]. **b** ORR/OER bifunctional LSV curve. **c** Discharge curves at different current densities. **d** Charge–discharge curves after frequent folding [170]

Furthermore, other Co-based SACs have also been constructed to realize flexible zinc-air batteries.

The development of central atoms, the synergy between multimetallic sites, and the design of supports are the strategies currently involved. Y. Wu et al. reported a high-temperature gas transport strategy to build single copper atoms on N-doped carbon with abundant defects [221]. Cu SAs with valences between 0 and $+ 2$ (Fig. 3.25a), complexed with 3 N atoms (Fig. 3.25b). The Cu ISAS/NC-based Zn-air battery achieved a power density of 280 mW cm^{-2} and a specific capacity of 736 mAh g^{-1} (Fig. 3.25c). Furthermore, Cu SAs (Cu-N$_3$) have been successfully immobilized on hollow carbon nitride-deficient nanospheres (CuSA@HNCN). Under the same synthesis conditions, the ORR/OER activity of CuSA@HNCN even exceeded that of FeSA@HNCN and CoSA@HNCN, illustrating the feasibility of Cu-based SACs for air electrodes (Fig. 3.25d) [222]. Mo$_1$N$_1$C$_2$ active sites achieved the lowest ORR and OER overpotentials compared to their nanoparticles and cluster counterparts (Fig. 3.25e) [223]. Theoretically, Patel et al. [224] chose a Cu-modified covalent triazine framework as a model SAC (Cu/CTF) and performed calculations. Based on electronic structure analysis and dissolution effects, an appropriate description for evaluating ORR activity is proposed, which not only uncovers the source of ORR activity on Cu/CTF, but can also be extended to other SACs.

By introducing multiple metal SAs simultaneously on carbon-based supports, the interaction between different metal atoms can enhance the catalytic performance. Chen et al. [225] prepared Fe and Co SA co-existing catalyst with an approximate MN$_4$ configuration. According to the free energy paths in theoretical calculations, Fe and Co atoms interact with each other in electronic structure, which is favorable

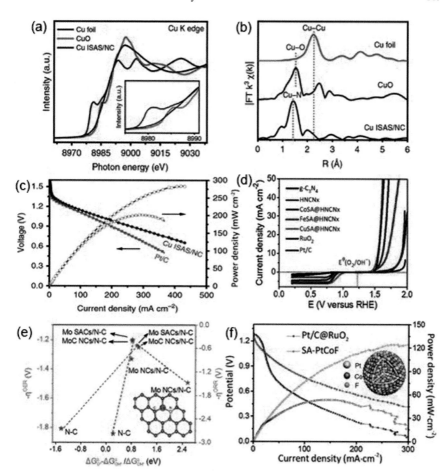

Fig. 3.25 a Copper K-edge XANES and **b** EXAFS results. **c** Discharge polarization curve and power density plot [205, 221]. **d** ORR/OER bifunctional LSV curve [222]. **e** The relationship between the theoretical overpotential of ORR/OER and the free energy of adsorption. The inset is the $Mo_1N_1C_2$ model [223]. **f** Discharge polarization curve and power density plot. *Illustration* SA-PtCoF model [227]

for the ORR process to occur. In the zinc-air battery, the FeCo-IA/NC-driven OCV is 1.472 V, and the power density is 115.6 mW cm^{-2}. Chen et al. [226] constructed hollow graphene nanospheres with Fe-N$_4$ sites in the outer layer and Ni-N$_4$ sites in the inner layer. Excellent ORR and OER properties derived from Fe-N$_4$ and Ni-N$_4$ species endow Zn-air batteries with high energy efficiency and durability. The metal-support interaction stabilizes the metal SAs on the support and modulates their geometric environment and electronic structure to enhance the intrinsic activity. The interaction between Co SAs, Co$_3$O$_4$ and N-doped carbon enhanced the activity of ORR/OER. Y. Yang et al. established platinum-cobalt alloy nanosheets with interstitial F doping (SA-PtCoF) [227]. The F atoms caused lattice disorder, weakened the platinum-cobalt bond, and finally produced a stable platinum SA at the edge of the nanosheets. The transfer of electrons from Co to adjacent Pt atoms pulls down the d-band center of Pt and tunes the adsorption behavior of the intermediate. Correspondingly, the electronic structure of Co is also affected by the neighboring Pt atoms, which can facilitate OH-adsorption during the OER process. The SA-CoF based zinc-air battery achieved a power density of 125 mW cm^{-2} (Fig. 3.25f). Here, Co supports also act as active species, while Pt SAs act as cocatalysts to some extent.

In short, Fe-N-C and Co-N-C sites are the most commonly used active species in cathode materials for Zn-air batteries. Although Fe-based SACs showed good ORR performance, Co-based SACs also presented distinct advantages in rechargeable batteries due to their bifunctional activity. Strategies such as heteroatom doping, multi-site coordination, support design, and increasing metal loading are effective.

For rechargeable ZABs, it is crucial to develop low-cost electrocatalysts for efficient ORR and OER. Pyrolysis of metal-containing complexes, MOFs or organic polymers is the most straightforward method to fabricate SACs for ORR and OER, which has been widely reported [44]. Careful selection of precursors and precise control of synthesis parameters are critical to achieving ideal single-atom sites. For example, Q. Zhang et al. reported a polymerization-decomposition strategy to synthesize atomically dispersed Fe-N$_4$ catalysts (Fig. 3.26a). Benefiting from predesigned bimetallic Zn/Fe polyphthalocyanines, highly dispersed single-atom metal sites can be achieved through low-temperature solvent-free solid-phase synthesis (Fig. 3.26b, c). ZAB using the obtained catalyst showed high stability at 2.0 mA cm^{-2} without obvious decay after 108 cycles (Fig. 3.26d) [199]. However, low metal atomic densities (usually below 5 wt% or 1 at%) limit their overall catalytic performance. Recently, H. Wang et al. synthesized a SAC with up to 40 wt% or 3.8 at% of transition metal atoms [228]. They used graphene quantum dots as supports, which provided many anchoring sites, avoiding aggregation and favoring the generation of high densities of transition metal atoms [132]. Furthermore, G. Wu et al. reported a two-step pyrolysis strategy involving doping and adsorption processes, effectively increasing the active material density (3.03 wt%) [229]. Although high-temperature pyrolysis is a facile method to prepare SACs, unpredictable structural changes in the support and the formed inorganic materials can affect their activity and stability due to attack by oxygen-free radicals. To address this issue, synthetic methods that do not require pyrolysis have been proposed to prepare M-N$_x$-coordinated SACs. However, it remains a great challenge to precisely control active centers with reversible oxygen reactions. MOF-based materials, especially ZIFs, have become one of the most popular precursors for the synthesis of ORR catalysts. Recently, Z. Chen et al. developed pristine Co-Zn heterometallic ZIFs without pyrolysis as an air electrode for ZABs [230]. The results show that oxygen electrocatalysis is promoted by generating unoccupied $3d$ orbitals at metal sites through a competitive coordination strategy. ZAB using the as-prepared ZIFs catalyst showed higher energy efficiency and excellent cyclability for 1250 h at 15 mA cm^{-2}. To demonstrate the role of metal centers in MOFs on ORR performance, Y. Peng et al. developed conducting coordination polymers by a simple ammonia-assisted procedure (Fig. 3.26e) [231]. Compared with

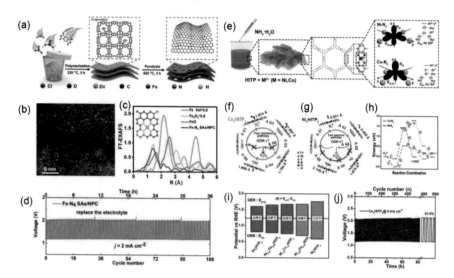

Fig. 3.26 **a** Synthesis process of Fe-N$_4$ SAs/NPC. **b** Distortion-corrected HAADF-STEM of Fe-N$_4$ SAs/NPC. **c** Fourier transform (FT) k3-weighted EXAFS. **d** Charge–discharge cycling performance of ZAB [199]. **e** Schematic diagram of the synthesis process of M$_3$HITP$_2$. **f** Catalytic cycling of Co$_3$HITP$_2$ and **g** Ni$_3$HITP$_2$ and bond lengths of ORR intermediates. **h** Free energy diagram of ORR on Ni-N$_4$ and Co-N$_4$. **i** Graph illustrating the difference between E$_{1/2}$ for ORR and E$_{j=10}$ for OER. **j** Electrostatic cycling performance of ZAB at 5 mA cm^{-2} [231]

the Ni coordination polymer (Ni$_3$HITP$_2$), the unpaired $3d$ electrons in the Co coordination polymer (Co$_3$HITP$_2$) have lower conductivity, which contributes to higher electrocatalytic activity in ORR and OER. DFT was then employed to reveal the source of the activity, which showed a transition from a two-electron pathway on Ni$_3$HITP$_2$ to a four-electron pathway on Co$_3$HITP$_2$ (Fig. 3.26f). ZAB with Co$_3$HITP$_2$ ran stably for over 80 h at 5 mA cm^{-2} (Fig. 3.26g, h).

In addition to the direct modulation of MOF-based materials, metal phthalocyanines and some phthalocyanine-based layered 2D conjugated MOFs assembled with carbon supports via intermolecular interactions have also been reported for the synthesis of SACs (Fig. 3.27a) [232]. Unlike the random creation of single atomic sites during pyrolysis, the obtained COF comes with pre-assembled Fe-N-C centers riveted directly onto the graphene scaffold. Electron localization function analysis revealed only van der Waals interactions. The Fe-C electron pathway is confirmed by the fact that graphene electrons are attracted to N-coordinated Fe sites by comparing the difference in charge density (Fig. 3.27b). The synthesized SAC showed higher ORR catalytic performance, and the ZAB based on the mixture of SAC and IrO$_2$ exhibited a long lifetime of over 300 h (Fig. 3.27c). X. Feng et al. developed a copper phthalocyanine-based 2D conjugated MOF supported on CNTs as an air electrode for ZAB (Fig. 3.27d) [233]. The obtained 2D conjugated MOFs show a highly crystalline structure with square planar cobalt bis(dihydroxy) complexes (Co-O$_4$)

as linkers (PcCu-O₈-Co) (Fig. 3.27e). In situ Raman spectroscopy-electrochemistry and DFT verify that the Co center is the catalytic site.

Generally speaking, O_2 molecules tend to adsorb on SACs in an end-pair model, while O_2 molecules tend to adsorb on metal clusters in a bridging model. Compared with end adsorption, bridge adsorption is more favorable for the 4e⁻ ORR pathway [61]. In view of this, the diatomic central site is considered an effective way to further enhance its catalytic activity [124, 134, 135]. Among these diatomic center site structures, the incorporation of double transition metal single atoms into M-N-C structures

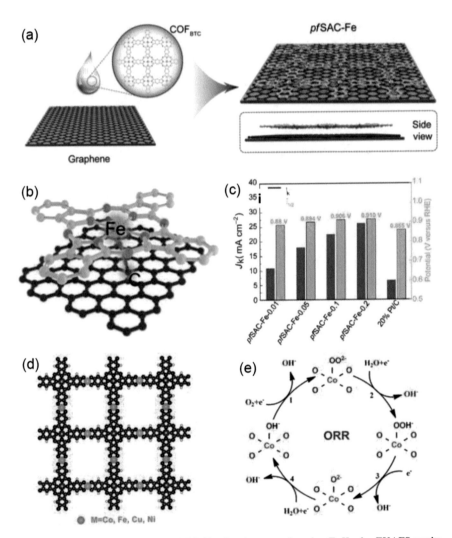

Fig. 3.27 **a** Synthesis of FePc-rich COF. **b** Simulated structure based on Fe K-edge EXAFS results. **c** Comparison of kinetic energy current density and half-wave potential [232]. **d** Schematic diagram of the structure of PcCu-O₈-M (M = Co, Fe, Ni, Cu). **e** Proposed reaction mechanism [233]

has been well studied, which may lead to synergistic effects between the two metal atoms due to charge redistribution and d-band center shift [136–138]. For example, the incorporation of Fe-Co bimetallic single atoms into the M-N-C structure, forming N-coordinated diatomic center points, helps to enhance ORR activity (Fig. 3.28a) [234]. Changes in the electronic structure and active site geometry will stretch the length of the O–O bond and weaken the binding. Besides the M-N_4 structure with bimetallic atoms, combining M-N_4 sites with only one metal atom is also beneficial to improving the electrocatalytic performance [140, 141]. H. Fu et al. prepared atomically dispersed Cu-N_4 and Zn-N_4 on N-doped carbon supports (Cu/Zn-NCs) [235]. DFT calculations indicated that the Cu-N_4 adsorbed on the end face acted as the main active center of the ORR. Due to the transfer of electrons from the Zn atom to the d orbital of the Cu atom, the electronegativity of the Cu center increases, favoring the absorption of intermediates and breaking the OO bond. In addition, certain metallic SACs can provide excellent electrolytic properties for specific reactions. Most SACs with a single catalytic function focus on ORR or OER processes, which cannot meet the requirements of rechargeable ZABs. Given that the ORR and OER activities of SACs with different transition metal centers follow a different order, Janus catalysts with single-atom selective combinations are expected to endow them with excellent bifunctional properties [143]. For example, T. Ma et al. developed a stepwise self-assembly strategy that enables Ni-N_4 and Fe-N_4 sites to be located on the inner and outer walls of graphene hollow nanospheres, respectively (Fig. 3.28b) [226]. DFT calculations show that the outer Fe-N_4 contributes greatly to the efficient ORR, while the inner Ni-N_4 clusters are responsible for the excellent OER (Fig. 3.28c).

Recently, the synergistic effect of single-atom central sites and clusters/nanoparticles has also been shown to enhance ORR activity [144, 145, 187]. For example, Z. Wang et al. reported an efficient and durable ORR catalyst for rechargeable ZABs [236], which consisted of atomically dispersed Co single atoms (Co-SA) and small Co nanoparticles (Co-SNPs). Their DFT calculations revealed that the rate-determining step (RDS) of the Co-N_4 system intermediate (*OH + e^- → *OH^-, represents the catalytically active site) limited the overall ORR reaction at the Co-N_4 site (Fig. 3.29a, b). In the case of Co-N_4@Co_{12} and Co-N_4@Co_2 layers, the energy barrier of RDS (*OOH + e^- → *O + OH^-) of Co-N_4@Co_{12} is lower, indicating a lower energy barrier and more efficient ORR catalytic activity than Co-N_4 sites. Compared with Co-N_4, the d-bands of active Co atoms in Co-N_4@Co_{12} and Co-N_4@Co_2 layers show a clear shift of the Co $3d$ state to lower energy levels (Fig. 3.29c). The interaction between Co-N_4 molecules and Co_{12} clusters is favorable for enhancing the intrinsic ORR catalytic activity compared with the optimized surface adsorption capacity of the intermediates. However, in the case of the Co-N_4@Co_2 layer, the peroxidation status of Co sites would lead to too weak adsorption capacity of intermediates, thus hindering the progress of ORR (Fig. 3.29d). Recently, S. Sun et al. synthesized and constructed a FeCo-N-C catalyst (M/FeCo-SA-N-C) containing highly active nanoparticles and M-N_4 recombination sites [237]. The corresponding DFT revealed that there was also a strong interaction between the M-NPs and FeN_4 sites, which can activate the O–O bond, thereby promoting the direct $4e^-$ process.

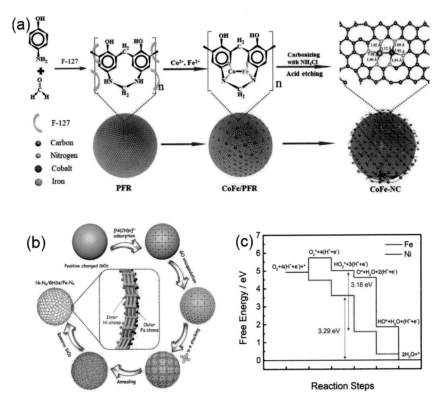

Fig. 3.28 a Synthesis procedure of CoFe-NC [234]. **b** Synthesis of SACs with double Ni-N$_4$ and Fe-N$_4$ sites. **c** Free energy diagram of the oxygen electrocatalytic reaction (U = 0 V) [226]

Although M-N-Cs coordinated by four N atoms have great advantages due to their attractive properties, the strong electronegativity of N atoms may inappropriately increase the adsorption free energy of reaction intermediates at metal centers. The Sabatier principle states that the best catalysts should bind O atoms and intermediates with the best strength. Too strong interaction is not conducive to the desorption of the product. To further optimize the local coordination environment of metal centers, the introduction of relatively weakly electronegativity secondary heteroatom dopants (S and P) as ambient atoms or coordination atoms has been investigated. For example, X. Feng et al. developed an efficient ORR of atomically dispersed nickel coordinated to nitrogen and sulfur species in porous carbon nanosheets (1.51 V at 10 mA cm^{-2} and Tafel slope of 45 mV dec^{-1}) [238]. Furthermore, single-atom copper catalysts with S atoms as ambient atoms for Cu-N$_4$ showed enhanced ORR performance [144]. Recently, P. Chen et al. synthesized a S-doped single-atom Co catalyst (CoSA/N, S-HCS) for efficient ORR and OER (Fig. 3.30a–c) [239]. CoSA/N, S-HCS exhibited superior ORR and OER catalytic activities in terms of half-wave potential (0.85 V) and overpotential (306 mV at 10 mA cm^{-2} current density) compared with the undoped samples (Fig. 3.30d, e). Furthermore, K. Müllen et al. developed Fe-N-Cs

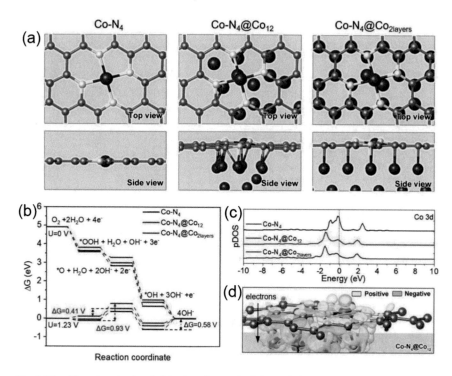

Fig. 3.29 **a** Top view (top) and side view (bottom) of the optimized atomic structures of Co-N₄, Co-N₄@Co₁₂ and Co-N₄@Co₂ layers. **b** ORR free energy diagram along the 4-electron pathway at U = 0 V and 1.23 V. **c** Calculated predicted density of states (PDOS) of the Co d band. **d** The charge density difference of Co-N₄@Co₁₂, the yellow and blue regions represent the accumulation and depletion of electrons, respectively [236]

with S and F doping as efficient ORR catalysts ($E_{1/2} = 0.91$ V) (Fig. 3.30f, g) [240]. The multi-layer stabilization strategy achieves high metal loadings ($\approx$16 wt%). DFT calculations indicate that OH* reduction is the RDS of the FeN₄ active site and the ORR process on Fe-SA-NSFC is thermodynamically favorable. In addition, the doping of S and F significantly reduced the OH* reduction free energy on the FeN₄ active site (Fig. 3.30h–j).

The local carbon structure surrounding the FeN₄ molecule plays a key role in the final catalytic activity. By tuning the coordination environment of nitrogen to obtain different FeN₄ structures, mainly including body and gable-side hosting, FeN₄ sites have been identified as an efficient way to tune the adsorption free energy of intermediates in the ORR/OER process [148]. The results suggest that the defective herringbone-edge FeN₄ configuration may be a highly active site. For example, micropores in M-N-Cs lead to the formation of more FeN₄ sites located at the edges, which exhibit higher ORR performance compared with FeN₄ sites on the plane. Meanwhile, the degree of graphitization of the carbon support plays a key role in determining the electrochemical stability, especially during the harsh OER process.

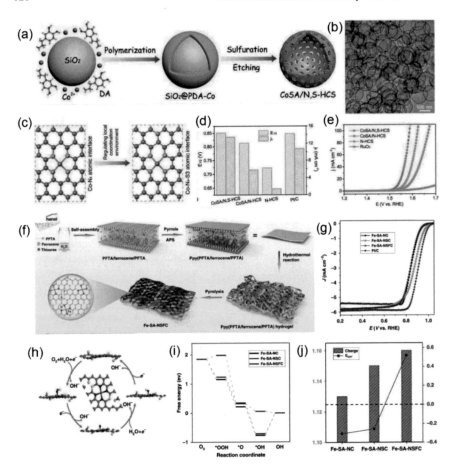

Fig. 3.30 **a** Schematic diagram of the synthesis process of CoSA/N, S-HCS. **b** TEM image of CoSA/N, S-HCS. **c** Atomic interface model of CoSA/N, S-HCS. **d** Comparison of half-wave potential ($E_{1/2}$) and kinetic energy current density (j_k). **e** OER polarization curve [239]. **f** Schematic diagram of the synthesis process of Fe-SA-NSFC. **g** ORR polarization curve. **h** Proposed mechanism of 4e⁻ ORR reduction on Fe-SA-NSFC. **i** Free energy diagram of the catalyst. **j** OH* binding energy and Mulliken charge of FeN₄ active sites in the catalyst [240]

Highly graphitized carbon nanotubes and graphene with strong oxidation resistance have been identified as ideal supports for single atoms. However, integrating more FeN₄ sites into highly graphitized carbon substrates to achieve bifunctionality while maintaining high catalytic activity is a major challenge. Recently, Z. Chen et al. developed a self-sacrificial templating method to integrate edge-enriched FeN₄ sites into highly graphitized carbon nanosheets [241]. The Fe clusters formed during the synthesis catalyzed the growth of graphitized carbon and induced FeN₄ to be preferentially fixed around the porous structure. The synthesized catalysts showed excellent catalytic activity and stability in terms of ORR (half-wave potential of 0.89

V) and OER (overpotential of 370 mV at 10 mA cm^{-2}). The assembled rechargeable ZABs exhibited high energy efficiency and enhanced cycling stability, lasting more than 240 cycles. X. Liu et al. developed cobalt single atoms anchored on nitrogen-doped graphene-sheet@tube (CoSAs-NGST) to study its morphological effect on the surface electrical structure (Fig. 3.31a) [242]. The hybrid structure of bamboo-like graphene tubes and sheets enhanced the dispersion of single atoms and induced the evolution of defect states. DFT modeling indicated that the coupling effect of Co-N$_4$-tubes and Co-N$_4$-sheets contributed to enhanced ORR and OER activities (Fig. 3.31b–d). The excellent bifunctional catalytic performance of CoSAs-NGST exhibited more robust stability in ZABs and presented a small voltage gap of 0.93 V at 5 mA cm^{-2}. To date, most SACs with M-N-C molecules are based on carbonaceous substrates (amorphous carbon, graphene, carbon nanotubes, carbonitrides, etc.), and they generally show improvements in ORR. However, carbonaceous catalysts for ORR processes still suffer from insufficient activity and durability due to the slow oxidation of carbonaceous substrates at high potentials (>1.8 V) [151].

Recently, A. Wang et al. developed a durable and conductive tungsten carbide (WC$_x$) support [243] in which the catalytic sites of FeNi atoms are weakly bound to surface W and C atoms (Fig. 3.32a). They believed that the formation of heteroatom bonds of metal atoms, such as metal-N$_x$-C$_y$, may adversely affect the catalytic activity. Stabilizing catalytic metal atoms on the surface of WC$_x$ supports without the help of strong heteroatom coordination is expected to further enhance catalytic activity and durability. They found that these atomically dispersed FeNi atoms were supported on the surface of the WC$_x$ through metal–metal interactions (Fig. 3.32b, c). Furthermore, they used DFT calculations to elucidate the source of activity and the effect of surface oxidation on OER performance. The results show that WC$_x$-FeNi exhibits the lowest OER overpotential (0.16 V) compared to WC$_x$-Ni and WC$_x$-Fe catalysts (Fig. 3.32d, e). In addition, the synergistic effect of Fe and Ni in the O-bridged FeNi molecules formed during OER further enhanced the catalytic activity of OER (Fig. 3.32f, g).

3.3.5 Transition Metal Alloy Catalysts

Metal and metal alloy nanoparticles supported on carbon nanomaterials have been widely explored as oxygen electrocatalysts, while cobalt is the most frequently considered bifunctional ORR/OER active material. Fig. 3.33a, b show TEM images of nickel–cobalt alloy nanoparticles decorated on carbon nanorubber [244]. The fibers were pyrolyzed under the protection of N$_2$, and the metal ions were simultaneously reduced to metal alloys. The lattice pattern in Fig. 3.33b is close to that of the different faces of the metals nickel and cobalt. Catalysts with varying amounts of NiCo alloys were synthesized, and the ORR and OER properties in 0.1 M KOH are shown in Fig. 3.33c, d. The best electrocatalyst achieved an ORR current density of 3.0 mA cm^{-2} at a potential of 0.81 V (a half-wave potential of 0.80 V) and an OER current density of 10 mA cm^{-2} at a potential of 1.76 V. Before electrochemical

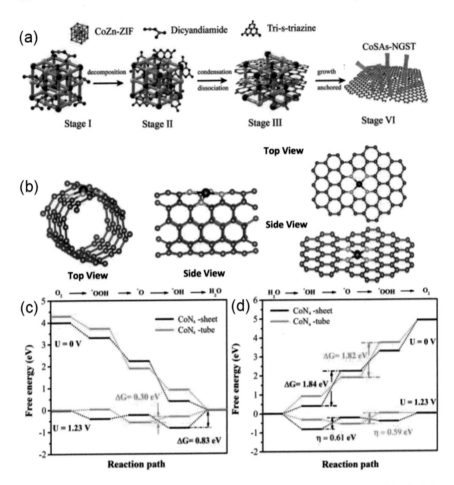

Fig. 3.31 **a** Schematic diagram of the synthesis process of CoSAs-NGST. **b** DFT models of tubular and sheet-like CoN₄. **c** Free energy diagrams of the ORR and **d** OER processes on both models [242]

testing, Co_2 and Ni_2 could be detected on the catalyst surface. At the same time, after electrocatalysis, the metal elements were oxidized to higher valence levels, indicating that metal (oxy)hydroxides provided the main activity of the catalyst (Fig. 3.33e, f).

Alternative explanations have also been proposed for the origin of the electrocatalytic activity of metals and alloys. T. Ma et al. reported encapsulation of bimetallic FeCo nanoparticles in nitrogen-doped carbon nanotubes as oxygen electrocatalysts [180]. CNTs were grown in situ on metal alloy nanoparticles with g-CN as carbon and nitrogen sources. The bifunctional electrocatalytic activity of the catalyst was tuned by varying the iron/cobalt ratio. However, after treatment with strong acid to remove metal species, the catalyst showed negligible activity loss. Iron-cobalt alloys

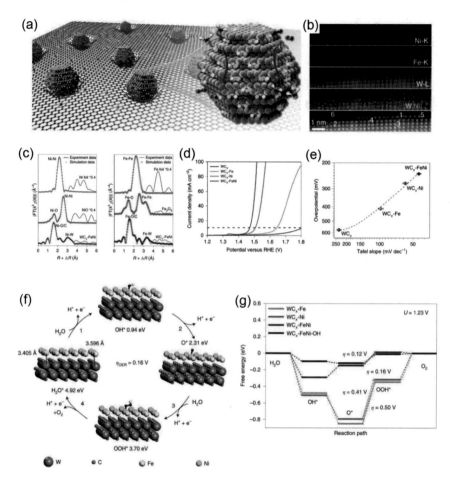

Fig. 3.32 **a** Schematic diagram of iron and nickel atoms stabilized on WC_x nanocrystals. **b** STEM spectral imaging of the edge of the nanocrystalline. **c** k3-weighted FT spectra of nickel K-edge and Fe K-edge. **d** OER polarization curves of different catalysts. **e** OER data analysis and Tafel slope at a current density of 10 mA cm^{-2}. **f** Reaction paths and corresponding free energies on WC_x-FeNi. **g** The reaction pathway of OER at a set potential of 1.23 V [243]

are believed to affect the morphology and electronic structure of N-doped carbon as the main electroactive site.

3.4 Carbon-Based Catalysts

Carbon-based electrocatalysts offer a promising route to develop efficient bifunctional ORR/OER electrocatalysts with chemical durability, excellent electrical

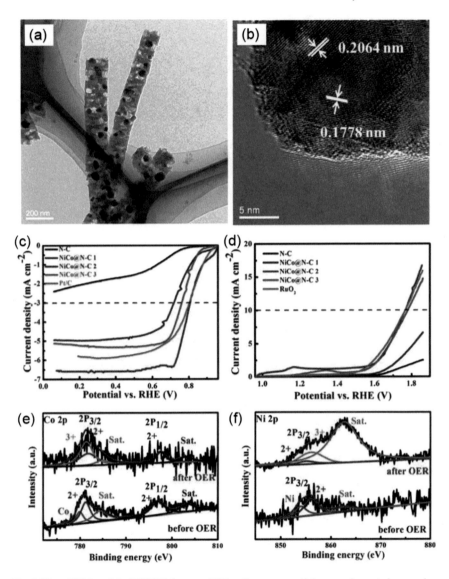

Fig. 3.33 a TEM and **b** HRTEM images, NiCo alloy nanoparticles are decorated on carbon nanofibers. **c** ORR and **d** OER LSV maps at a scan rate of 10 mV/s in O_2-saturated 0.1 M KOH solution. **e** Co 2p and **f** Ni 2p XPS spectra before and after OER electrocatalysis [244]

conductivity, well-developed structures, and controllable structures. Research on carbon-based materials has mainly focused on the following aspects: (1) increasing exposed active sites and constructing active networks; (2) improving mass transfer and reaction kinetics, using hierarchical porous structure optimization; (3) designing with high robust electrocatalysts with electrochemical stability. Carbon nanomaterials, including 0D (e.g., fullerenes), 1D (e.g., carbon nanotubes (CNTs)), 2D (e.g., graphene), and 3D porous carbon nanostructures, have been widely reported. Engineering to construct defective carbons, such as carbon modification, carbon electronic structure and surface chemical state skeleton, is an effective strategy to further refine their electrochemical performance. Specifically, the intrinsic carbon defects can be directly used as effective active sites. At the same time, their binding species with heteroatom dopants/metals can also be used to construct other effective synergistic active networks; due to the improved overall conductivity and better coupling synergy effect, the hybrids exhibit excellent electrocatalytic activity and stability. Furthermore, the porous structure of defect-rich carbon materials can reduce electron transfer resistance, accelerate oxygen and electrolyte diffusion, and expose more active sites.

3.4.1 Undoped Carbon-Based Catalysts

Non-metallic carbon-based materials have been widely studied as ORR catalysts, however, there are few reports on pure carbon bifunctional catalysts. The main reason is that the OER active sites of carbon materials are insufficient, and carbon is easily oxidized or corroded during the OER process.

The reaction process of functional oxygen electrode involves the steps of adsorption, reaction and desorption of reactants at the active site. Pure carbon-based materials can meet the requirements for electron transfer in catalytic reactions and the rapid diffusion and transfer of reactant molecules and electrolytes. Still, they cannot provide enough active sites for catalytic reactions and cannot fundamentally solve the problem of carbon inertness to the reaction itself. To advance the application of carbon-based materials in electrocatalysis, their activation modification is crucial.

Improving the oxygen-functional catalytic performance of carbon-based materials can mainly be done from two aspects: increasing the active sites of the catalyst and promoting the full exposure of the active sites. The former is mainly achieved through active material modification, heteroatom doping, and defect construction, while the latter mainly includes enhanced conductivity and porosity. Ideally, the two complement each other.

Modification by active substances: Grafting metals with high catalytic activity and their metal compounds on carbon materials can effectively improve the catalytic activity. Among them, metal compounds mainly include metal oxides, sulfides, alloys, etc.

Heteroatom doping: It is a mainstream strategy to improve carbon-based materials. After heteroatoms enter the carbon skeleton in the form of doping, they can adjust the electronic state density of carbon atoms, generate active centers, and improve catalytic performance. Heteroatoms mainly include N, B, P, S and the like.

Porosity: Rich porosity and large specific surface area are conducive to fully exposing the active sites on the catalyst surface, increasing the electrochemically active area, and promoting the transport of reactive species. At present, the preparation methods of porous carbon materials are relatively mature, which can be realized by template method, high temperature pyrolysis, chemical vapor deposition, etc.

Conductivity: In the electrochemical reaction process, the good conductivity of the electrode is the premise, and the high conductivity is conducive to the transfer of electrons and improves the catalytic efficiency. However, not all carbon materials have good electrical conductivity. Generally, amorphous carbon materials have poor electrical conductivity. Therefore, when choosing amorphous carbon as a substrate, it is necessary to first consider improving its electrical conductivity.

3.4.1.1 1D CNTs

Due to its one-dimensional properties and good elasticity, the fibrous structure improves the space utilization of the nanofibers. ZABs with fiber-structured air cathodes provide a new direction for efficient and flexible electronics. Carbon nanofiber (CNF) and carbon nanotube (CNT) based materials have been used as efficient air electrodes. Y. Xu et al. developed a flexible and stretchable ZAB cathode by designing RuO_2/CNT sheets with aligned, cross-stacked, and porous structures [245]. The excellent electrochemical performance of this air cathode enables ZAB to show stable discharge voltage at high current density (2 A g^{-1}). Afterward, Z. Guo et al. synthesized ordered mesoporous RuO_2/CNF arrays using the natural crab shell template method [246]. Notably, the macroscopic voids between CNFs can facilitate oxygen transfer, while the improvement of ZAB shows a low overpotential and long cycle life. The uniform ruthenium oxide coating layer and the ordered mesoporous structure have good electrical conductivity and efficient ion diffusion properties, respectively. Recent studies have shown that array-structured carbon fiber GDE, a porous CNF cage for supercapacitor electrode scaffolds, carbon nanotube-grafted carbon polyhedral catalysts and other one-dimensional fibrous structure coupling materials had great potential for electrochemical performance.

3.4.1.2 2D Graphene

The advantage of the layered structure is that it can further enhance the performance of the battery by increasing the surface area and catalytic activity of the battery through a simple fabrication process and post-synthesis treatments such as exfoliation or exfoliation. In addition, the nano-thickness can provide layers with flexibility. In addition, it is clear that the 2D structure is conducive to the fast transport of electrons and ions. In layered carbon, the active catalytic components can be arranged in different ways, such as horizontal or vertically scattered on a plane, assembled into clusters, or completely filled in the spaces between layers. Sun et al. fabricated a hierarchical pore-distributed sandwich-like carbon black and cobalt nanocomposite using reduced graphene oxide (rGO) nanosheets, in which dispersed nanoparticles and isolated rGO nanosheets jointly enhanced the electrical conductivity of the electrode sheet rate, without aggregation, greatly improves the performance of ZAB. Z. Chen et al. employed a soft templating method to in situ assembled Fe_2N nanoparticles on N-doped graphene-like carbon surfaces [247]. The layered carbon capping layer formed on Fe_2N enhances the connection of nanoclusters to graphene and promotes the formation of Fe-N-C active sites, thereby promoting the performance of ZAB.

3.4.1.3 3D Porous Carbon

Porous structures have been widely used in carbon-based catalysts and air cathode catalysts to facilitate the transport of reactants and electrolytes. Carbon substrates with porous structures can be easily obtained by a top-down approach for optimal performance. H. Wang et al. applied a simple in-situ H_2 etching method to carbon cloth to achieve coaxial cable-like structures with nanostructured porous graphene skins [248]. The fabricated carbon cloth exhibits higher OER and ORR current densities than pristine carbon cloth, and can be used as an integrated air cathode for rechargeable flexible ZABs, which can maintain stable charge–discharge cycles even when bent. In terms of self-designed materials, Q. Liu et al. utilized electrospun polyimide pyrolysis to synthesize flexible nanoporous carbon fiber films, which exhibited excellent dual electrocatalytic activity, enabling flexible rechargeable ZABs with high round-trip efficiency and mechanical stability [249]. In addition, carbon-containing frameworks such as molecular sieves and MOFs are also frequently used to construct porous structures to utilize the pore volume fully. As for bottom-up strategies, template-assisted methods are usually employed to tune the pore structure. Salt template units have been successfully used as an inexpensive and safe material. Using NaCl as a template, Y. Chen et al. established a 3D foam-like porous carbon matrix during carbonization and found that the amount of NaCl versus the surface area plays a key role [250].

3.4.2 Heterogeneous Atom-Doped Carbon-Based Catalysts

The doping of heteroatoms (N, P, B, S, F, etc.) into the carbon matrix has proven to be a promising strategy as it modulates the surface chemistry, electronic configuration and adsorption/desorption tendency of oxygen intermediates, ultimately promoting OER and ORR kinetics. It has been reported that heteroatom doping in carbon matrix can inject non-electrically neutral states and active sites. Indeed, doping heteroatoms into the carbon lattice creates a difference in electronegativity, leading to asymmetry in charge distribution and electron spin density, giving carbon its metallic properties.

For example, doping carbon atoms with larger electronegative atoms N (3.0) or F(3.98) results in a net positive charge on adjacent carbon atoms, which is considered a favorable feature of O adsorption. As the most electronegative atom, F may also cause charge delocalization on carbon atoms, causing changes in charge density. N is usually the most utilized heteroatom dopant because N has 5 valence electrons, and the atomic size of N is similar to that of C atoms, resulting in the smallest lattice mismatch and strong covalent bonds. On the other hand, the electronegativity of dopant atoms such as B (2.0) is close to that of C (2.55) atoms. Then they form positively charged active sites (B + , C +), confirmed by DFT simulations. In this way, B becomes an active site, promoting the chemisorption of oxygen molecules on itself. It is found that B-doped carbons exhibit ORR activity comparable to that of Pt noble metal-based catalysts. This phenomenon of B-doped carbon may be related to the lack of electrons in B atoms. In the case of S doping, its electronegativity is almost close to that of C atom (2.58), and the atomic size is larger. Therefore, no charge transfer occurs for charge redistribution due to poor electronegativity. In contrast, S-doped C atoms with high electron spin density were found to be responsible for the ORR activity. For phosphorus doping, P (2.1) is less electronegative than carbon atoms and has a larger atomic radius, which modulates the charge density through the lone pair electrons on the 3p orbital and the lone pair electrons on the empty 3d orbital of oxygen. J. Zhang et al. proposed that P doping into the carbon lattice suppresses the reaction free energy barrier by creating a defect-induced active surface, favoring oxygen adsorption [251].

Recently, H. Yu et al. discovered that some doped heteroatoms were not necessarily bound to the carbon matrix or the bulk; instead, they may create some favorable environments for surface modification [252]. For example, some P heteroatoms are bonded to the N site in P, N double-doped carbon, but the P atoms are not integrated into the carbon matrix. Instead, these P atoms decorate the surface of the carbon material, creating a favorable environment for post-processing. Therefore, they subsequently treated $NaBH_4$ to form a pyridine-N-B(OH)$_3$ site. These sites are believed to be favorable for the adsorption of water molecules in the semi-free state. Therefore, this surface modification strategy accelerates the steps involving an attack by water molecules, leading to an increase in ORR activity. Oxygen doping, when O atoms and C atoms are doped, directly bond with $sp2$-hybridized carbon atoms. For oxygen doping, when O atoms are doped with C atoms, they directly bond with sp^2-hybridized carbon atoms. The intrinsic bands and electronic structures of the carbon matrix were subsequently regulated, increasing the specific surface area of the carbon matrix accompanied by multiple active sites.

3.4.2.1 Single Heteroatom Doping

L. Dai et al. reported similar work on N-doped carbon nanotubes (NCNTs) in 2009, which led to the extensive exploration of N-doped carbon materials. N-doped carbon materials have been most studied as a promising emerging catalyst for ORR and OER [253–255]. It is generally believed that pyridyl N, pyrrolyl N, and graphitic N play important roles in reversible ORR/OER. J. Li et al. reported N-doped ultra-thin carbon nanosheets (NCNs) with ultra-high specific surface area obtained by simply pyrolyzing a mixture of citric acid and NH_4Cl [256]. Insertion of N atoms at high pyrolysis temperatures leads to the formation of more edge and topological defects in NCNs. To further elucidate the potential effects of doping types and doping concentrations on ORR/OER catalysis, two types of structural models (14 different models) were constructed. Volcanoes of ORR and OER were obtained by plotting overpotential as a function of ~G(*O) on various possible active sites on N-doped graphene monolayers and armchair and chevron graphene nanoribbons picture. DFT calculations revealed that the intrinsic active sites for ORR and OER were carbon atoms located at the armchair edge and adjacent to graphitic N. Notably, linear swept voltammetry (LSV) curves showed the performance as a bifunctional oxygen electrode catalyst for OER and ORR. Furthermore, ZAB exhibited high energy density (806 Wh kg^{-1}), low charge–discharge voltage gap (0.77 V), and long-cycle stability. The relationship between the electrocatalytic activity of carbon materials and defects/dopants has also been quantified [257]. Electrochemical experiments and DFT calculations showed that the electrocatalytic activity of carbon was improved with the increase of defect sites and doping of pyridine N, and the electrocatalytic activity of the chevron-shaped and handrail-shaped carbon atoms at the edge was higher than that of the basal surface carbon atoms. However, carbon atoms surrounding pyridine N showed higher electrocatalytic activity than carbon atoms adjacent to graphitic N. This indicates that both edge defects and successful doping of pyridine N are beneficial to enhancing the electrocatalytic activity of carbon-based materials. In addition, topological defects and high content of desired pyridine N species in highly curved graphene sheets are also the keys to improving their electrochemical activity and stability [258].

H. B. Yang et al. provided the first experimental evidence that different N types in N-doped carbon nanomaterials acted on different electrocatalytic reactions [259]. N-doped graphene nanoribbons (N-GRWs) with interconnected three-dimensional (3D) structures were obtained by two-step carbonization of a mixture of melamine and L-cysteine under an argon atmosphere (Fig. 3.34a). Experimental data suggest that electron-donating graphitic N acts as the most active catalytic site for ORR. Likewise, the electron-withdrawing pyridine N as the active site is favorable for ORR. The isolated active sites of ORR and OER are located in the n-type (quaternary ammonium salt/pyrroline-N) and p-type (pyridine-N) domains, respectively (Fig. 3.34b). Independent optimization of these active sites prevents their cross-deactivation to ensure the stability and durability of N-doped graphene catalysts. N-GRW showed a slight negative half-wave potential shift and a small finite current change after continuous cycling after 2000 consecutive cycles (Fig. 3.34c), indicating that it has electrocatalytic operational activity and stability comparable to Pt/C. A similar situation occurred in the OER performance test (Fig. 3.34d), where N-GRW showed durability. The ZABs assembled from N-GRW showed cycling stability and could be rapidly charged and discharged for 150 cycles at 2 mA cm^{-2} (Fig. 3.34e).

Besides N doping, other heteroatom-doped carbon materials are also frequently studied [260]. Electronegativity of doped P-group elements (as electron donors) and their compounds (B, P, S, Si, Se, Sb, F, Cl, Br, I, POH, SOH, PO_2, SeO_2, and SO_2, etc.) and electron affinity can effectively change the electronic structure of carbon nanomaterials and promote ORR and OER, which are generally studied by

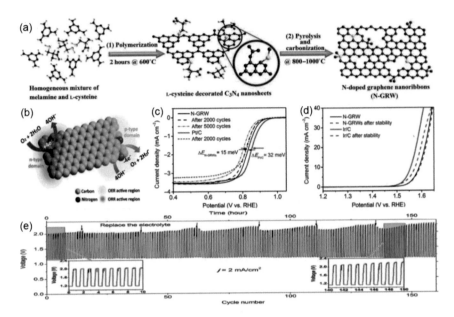

Fig. 3.34 a Synthesis steps of metal-free three-dimensional (3D) graphene nanoribbon network (N-GRW). **b** Schematic illustration of the ORR and OER occurring at different active sites of the n-type and p-type domains of the NGRW catalyst. **c** LSV curves of N-GRW and Pt/C before and after ADT at a scan rate of 50 mV s in 1 M KOH. **d** LSV curves of OER before and after stability testing of Ir/C and N-GRW. **e** Charge/discharge cycling at a current density of 2 mA cm^{-2}. The inset shows the charge/discharge curves of ZABs assembled with N-GRW as air catalyst after initial and long-term cycling tests [259]

trial and error methods. A recent DFT study identified an activity descriptor for the first time to predict bifunctional ORR/OER activity by establishing a volcano relationship between the intrinsic bifunctional activity of heteroatom-doped carbon-based catalysts and the descriptor [261].

Schematic representation of the possible positions of heteroatom dopants in graphene nanocarbons reveals the effect of changing the doping position in each structure. The active catalytic center in all cases was identified as carbon near the dopant. Elements such as B, P, Si, and Sb can act as ORR active centers; however, dopants N, S, Se, and halogens cannot. Experiments show that the edge of graphene has a faster electron transfer rate and higher electrocatalytic activity than the basal plane of graphene, which is the same as the effect of defect position on the catalytic activity. Likewise, the most active OER and ORR centers are also located near the edge of graphene, proving that incorporating P group elements near the edge is an effective way to enhance the OER/ORR activity of carbon materials. The minimum ORR/OER overpotential of heteroatom-doped graphene is shown as a function of the descriptor Φ. N was identified as the best dopant for ORR in graphene, while P exhibited the lowest OER overpotential. This suggests that the activity of doped carbon can surpass that of Pt and RuO$_2$, and that N- and P-doped carbons have the potential to be outstanding catalysts for ORR and OER, respectively. There are differences in charge density distribution between B, N, and F doped and undoped graphene sheets. Interestingly, in the same N-doped structure, two most active ORR and OER catalytic centers were found near the dopant. However, when F element was incorporated, the ORR/OER catalytic center appeared at the same site.

Zhang et al. fabricated PANI@CNTs composites by simply wrapping polyaniline (PANI) on CNTs. Except for self-doping N, there is no need to deliberately add additional heteroatom doping in the carbon matrix. The heterogeneous interfacial interaction between PANI and the encapsulated CNTs resulted in the exposure of abundant active sites in the CNTs, leading to elevated OER activity of the composite. In the assembled Zn-air battery, the PANI@CNTs composite exhibits large energy density and good Zn-air battery bifunctional activity within 36 h, so the performance of the PANI@CNTs composite is reasonable but not long-term stable. Similarly, K. Sheng et al. fabricated N-doped ultrathin carbon hollow spheres (5rG@NHCS) by encapsulating SiO_2 templates with polydopamine (N precursor), adsorbing graphene oxide, followed by carbonization and etching. 5rG@NHCS demonstrated its excellent discharge performance and confirmed that 5rG@NHCS could be used as a cathode for metal-air batteries in alkaline and neutral media [262].

L. Ma et al. prepared N-doped porous carbon (sp^2/sp^3 carbon interface) by pyrolysis of microalgae and ionic liquid [Bmim][FeCl$_4$]. Fe species played an auxiliary role in the formation of the sp^2/sp^3 carbon-based interface [263]. The synthesized N/biochar-800-7 exhibits excellent ORR performance, high CO and CH_3OH tolerance, and good capacity and stability. The ORR active site is pyridine-type N, not graphitic-type N. Similarly, Ilnicka et al. developed exfoliation of several layers of graphene followed by dipping in a suspension of Chlorella Vulgaris as a source of pyrrole-N and pyrrole-N functional groups into the carbon matrix. However, the obtained catalysts exhibited poor ORR activity in terms of onset potential, electron transfer number, and half-wave potential. Moreover, the OER is also much lower than that of contemporary electrocatalysts. Also, the battery performance is not that outstanding. X. Hao et al. employed melamine as an additional nitrogen source while utilizing organic waste green corn stover as a carbon precursor to enhance the electrocatalytic activity [264]. In addition, the activation of organic materials with FeCl$_3$ induced the recombination of corn stover biomass. FeCl$_3$ helped to increase the degree of graphitization of carbon to generate graphitic-N, which was crucial for improving electrocatalytic activity. Finally, the final product was etched with HCl to wash out FeCl$_3$. The results show that the ORR performance of the nitrogen-doped porous carbon material-supported Zn-air battery is better than that of other biomass carbon materials, with an onset potential of 0.985 V versus RHE, a peak power density of 127.9 mW cm^{-2}, and a specific discharge capacity of 794 mAh g$_{Zn}^{-1}$.

Tofu is another biomass that can be converted to doped carbon. However, carbon produced by direct pyrolysis of tofu suffers from low nitrogen content and small surface area. Therefore, H. Zheng et al. used urea-impregnated tofu and NaCl-assisted pyrolysis to protect N content and deliver high surface area. NaCl has a dual role: as a source of pore formation; and as a nanoreactor to protect the graded morphology and N content [265]. In addition, nitrogen content, especially reactive nitrogen species, such as pyridine nitrogen and graphitized nitrogen increases. Ultimately, the ORR activity of H-NHPC is superior to that of benchmark Pt/C catalysts and other biomass-derived N-doped carbon materials. Similarly, J. Cai et al. also employed a molten salt strategy to prepare nitrogen-doped carbon aerogels (NDC-MS) similar to the use of saline/optimally selected eutectic mixtures in glycine biomass as nitrogen source and carbon and molten salt as auxiliary solvent [266]. Glycine (NH_2-CH_2-COOH) is a source of abundant nitrogen-doped carbon with a large N/C atomic ratio. And $ZnCl_2$ with high vapor pressure plays a certain role in the pyrolysis process of organic precursors. The aerogel-like structure has a high surface area of 1549 m^2 g^{-1}. The higher the I_D/I_G value (1.81) of NDC-MS, the lower the graphitization degree of carbon and the more edge and structural defects caused by the molten salt template. Higher defects are believed to enhance ORR activity. Subsequently, NDC-MS was found to be a competitive ORR catalyst with reasonable half-wave potential, small H_2O_2 yield, excellent 4-electron channel, and large kinetic current density. In addition, NDC-MS also has a high peak power density (174 mV cm^{-2}). In general, noble metal-based (Pt/C and IrO_2) batteries lack the ability to operate in long-term cycling compared to N-doped carbons. Therefore, it is urgent to reveal the main performance indicators of the battery by monitoring the in-situ chemical changes inside the battery to ensure its long-term stable operation. To this end, H. Ji et al. performed in-situ XRD under real-time battery testing conditions to capture the comprehensive chemical transitions of N-doped carbon or Pt/C and IrO_2 integrated Zn-air batteries [267]. Two consecutive diffraction peaks exist at 44° and 54°, attributed to carbon, which are prominent throughout the operation. Indeed, N dopants have higher electronegativity, forcing a shortage of electrons in adjacent carbon atoms, making them more favorable for ORR active sites. In addition, after charging, the XRD diffractogram returns, ensuring it is reversible and long-term operation without undesirable by-products. On the other hand, (Pt/C and IrO_2) cannot exhibit this property, leading to their instability. Therefore, the optimal catalyst exhibited a low potential gap of 0.78 V and stable zinc-air performance over 400 cycles. Q. Lv et al. prepared pyridine N-mounted graphdiyne benzene ring (PyN-GDY) thin films by a modified Glaser-Hay coupling reaction to polymerize pentavinylpyridine monomers on a pyridyl-based copper foil substrate [268]. The Gibbs free energy diagram further confirmed that hydrogenation of adsorbed O_2 molecules was the rate-determining step of ORR, i.e., the formation of the *OOH intermediate. Furthermore, the acetylene carbon atom closest to the N atom is the most likely active site for P_yN-GDY. P_yN-GDY has excellent ORR performance for three main reasons: the high content of pyridine N atoms and the carbon atoms close to pyridine N exhibit Lewis alkalinity, which is the active site for oxygen adsorption and favors ORR activity; the large π-conjugated configuration is considered to be a favorable feature for ORR electron transport; and the three different types of macropores such as triangular, hexagonal and tetragonal pores facilitate convenient mass transfer.

Carbon materials have the remarkable advantages of low cost, good stability, large specific surface area and high electrical conductivity. They have attracted extensive attention as catalyst supports or non-precious metal catalysts. In the electrochemical catalysis of oxygen, oxygen can be adsorbed and reduced on the surface of carbon materials, but the oxygen reduction reaction on the surface of carbon materials is carried out in a two-electron way to generate peroxides, which often requires a high overpotential and is not conducive to increase the energy utilization efficiency. To improve the oxygen reduction performance of carbon materials, the catalytic activity can be enhanced by non-metallic heteroatom (N, S, P, B) doping or structural tuning [190]. Gong et al. [269] first identified non-metallic carbon materials as effective ORR electrocatalysts and showed that doping nitrogen in carbon nanotubes changed the chemisorption pattern of oxygen and promoted the four-electron reaction, which could effectively weaken or break the O–O bond, significantly improving the ORR catalytic activity. Nitrogen doped in carbon materials exists in the form of pyridine-type nitrogen, pyrrolic-type nitrogen and graphitic-type nitrogen, and three forms of different proportions can be obtained by changing the nitrogen source, adjusting the temperature and the synthesis method, but there is no specific conclusion on the form of nitrogen affecting the catalytic activity [270].

H. B. Yang et al. prepared a highly active nitrogen-doped three-dimensional graphene nanoribbon network (N-GRW) by pyrolyzing a mixture of melamine and L-cysteine and controlling the carbonization temperature [259], which exhibited an open circuit voltage of 1.46 V, a peak power density of 65 mW cm^{-2}, a specific capacity of 873 mAh g^{-1}, as well as good cycling stability in a zinc-air battery. The analysis suggests that the pyridine nitrogen structure plays the main catalytic role. In order to increase the content of pyridine nitrogen, S. Yi et al. used potassium permanganate to etch hollow tubular polypyrrole to prepare nitrogen-doped carbon nanotubes. By etching the carbon material, rich pore structures and defects were introduced, so that more N atoms were located at the edge position. The content of pyridine nitrogen increased from 32.8% to 45.2% (Fig. 3.35a) [271]. The catalyst had better half-wave potential (0.83 V) and stability than commercial Pt/C under alkaline conditions. A zinc-air battery showed a peak power density of 122 mW cm^{-2} and discharge specific capacity of 835 W h kg^{-1} (Fig. 3.35b).

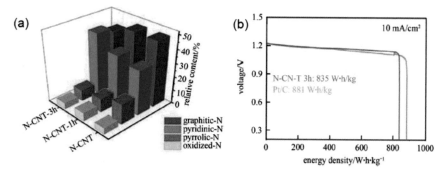

Fig. 3.35 **a** Distribution of graphitic-N, pyrrolic-N, pyridinic-N, and Oxidized-N obtained from the N 1 s spectra of as-prepared catalysts; **b** Energy densities of Zn-air batteries with N-CNT-3 h and Pt/C as air cathode catalyst [271]

The synergistic effect of different atom-doped carbon materials can improve the catalytic performance. Wu et al. prepared N, F, and P ternary doped macroporous carbon fibers (NFPC) by electrospinning [272]. The prepared NFPC catalyst had good electrical conductivity, large specific surface area and porous structure. The synergistic effect of heteroatoms showed better bifunctional catalytic activity for ORR and OER. In addition, nitrogen-doped carbon materials prepared by other methods such as hard template method [273] and CVD method [213] also have excellent electrochemical performance.

3.4.2.2 Multiple Doping

In addition, introducing multiple heteroatoms into nanocarbon electrocatalysts can effectively tune the ratio and electronic structure of doping atoms, making their catalytic activity higher than single heteroatom doping. The catalytic synergistic effect generated by the coexistence of several heteroatoms has been shown to further enhance the catalytic performance of carbon materials for ORR/OER [260, 274–276].

To date, N, P co-doped carbons have been widely reported as efficient bifunctional catalysts for ORR and OER [251, 277, 278]. To clarify the effect of doping type and doping position on the ORR/OER catalytic activity of N, P co-doped carbons, all possible doping structure types, such as isolated N-doped, isolated P-doped and/or N, P-coupled doping were constructed. DFT calculations showed that the synergistic effect of N and P co-doping yielded electrocatalytic OER and ORR activities superior to isolated N doping and isolated P doping [251]. Interestingly, recent DFT calculations with active sites located at the edge of graphene showed that N and P doping-induced charge transfer effectively tuned the electronic properties of carbon near the doping site, which was beneficial for optimization and creation of new active sites [237]. The doping interaction of N and P and the high level of pyridine N in carbon materials can expose more intrinsic defects as bifunctional electrocatalytic active sites. Among them, edge doping is more effective for enhancing catalytic activity under the condition of lower overpotential [278]. N, S co-doping of carbon is also frequently reported [279, 280]. It is found that N, S co-doping can greatly enhance the electrocatalytic activity of pyridinic N and graphitic N, and more defects are found to provide more active sites, further enhancing the catalytic activity of OER. The performance of N, S co-doped carbon as an air cathode for ZABs is close to that of the current state-of-the-art Pt/C-RuO$_2$, indicating that N, S co-doped carbon can serve as an efficient bifunctional catalyst for ZABs [279]. C. Zhi et al. reported a novel and facile strategy for the construction of heteroatom-enriched porous carbon catalysts via a one-pot pyrolysis reaction [281] (Fig. 3.36). Electron energy loss spectroscopy (EELS) elemental mapping confirms that N and S doping are uniformly distributed in the carbon framework. The ΔE ($E_{j=10} - E_{1/2}$) value of N, S-enriched hierarchical porous carbon in 0.1 M KOH is 0.81 V, which is superior to Pt/C and IrO$_2$, and even many metal-free and TM-based bifunctional catalysts. The performance of the optimized ZABs is significantly better than that of

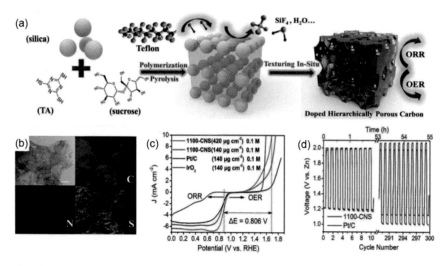

Fig. 3.36 **a** Illustration of the one-pot fabrication process of N, S-rich hierarchical porous carbon. **b** EELS spectra of different elements in the N, S enriched hierarchical porous carbon. **c** LSV curves showing the bifunctional ORR/OER activity of different samples in 0.1 M KOH. **d** Cycling test of the changeability of ZABs at 10 mA cm^{-2} using an N, S-rich layered porous carbon sample or a Pt/C sample as the catalyst [281]

the Pt/C-based ZABs. In addition, N, B co-doped carbon, B and C co-doped carbon nanotubes, and N, F co-doped graphene have also been reported [282].

Taking advantage of the synergistic effect, a special strategy for ternary heteroatom doping in carbon was also investigated. C. H. Choi et al. developed ternary heteroatom-doped carbon as an ORR catalyst [283], and the results showed that (B, P, and N) ternary doped carbon had significantly improved performance over single-atom doping. Tri-doped graphene-based materials, such as N-P-S and N-S-F, have also been synthesized and demonstrated to have higher catalytic activities; however, the understanding of the relevant mechanisms is largely lacking. Woo has reported an N, B and P co-doped carbon material. He proposed a synergistic way in which P improved charge delocalization and generated massive edge defects, while B increased the Pyridinic-N moiety and strengthened the sp2 framework. P, N and S ternary doped carbons are designed with electrostatic components. XPS analysis detected thiophene-S and 3 types of N, especially in addition to P–C bonds, where P was bonded to N. All PNS-PCs had poor crystallinity, and PNS-PC-2 with higher P content even exhibited amorphous properties, indicating a large number of internal defects [284–286].

The introduction of heteroatoms can tune the charge distribution and electronic properties of carbon, resulting in abundant active sites. The heteroatom-doped active sites greatly enhance the ORR/OER performance of carbon materials. However, carbon corrosion that occurs at high oxidation potentials (OER processes) limits their large-scale application in rechargeable ZABs because it leads to increased charge and mass transport resistance, and degradation of ORR/OER activity. So far, the development of heteroatom-doped carbon materials has not been able to meet the requirements of ZABs for practical applications, especially working at large current densities [282, 287]. A relatively high degree of graphitization is crucial to improving the corrosion resistance of carbon-based catalysts. Therefore, it is necessary to develop heteroatom-doped carbon materials with high corrosion resistance, such as N-doped carbon nanotubes (NCNTs) [229]. Furthermore, by developing strongly coupled TMs/nanocarbon hybrid materials, their electrocatalytic activity and stability can be improved.

3.4.3 Transition Metal-Modified Carbon-Based Catalysts

Introducing metal elements into carbon-based nanomaterials is another effective way to improve electrocatalytic activity. Carbon nanomaterials can promote the dispersion and utilization of effective catalytic sites, thereby enhancing catalytic activity. Heteroatom-doped carbon materials possess high specific surface area, high electrical conductivity, and sufficient functional groups to provide intimate interactions with metal-based materials. This interaction can effectively suppress the cathodic corrosion process, resulting in catalysts with high durability and chemical stability TM (e.g., Fe, Co, Ni, Cu, Mo, and W) and nitrogen-co-doped carbon-based (M-N-C)-based or carbon-based catalysts have been shown to be one of the most promising bifunctional candidates for OER and ORR. Intrinsic carbon defect sites, such as vacancies or topological defects, can also provide unique sites for trapping metal species. The large difference in electronegativity between TMs and carbon atoms produces sufficient charge transfer to make these building blocks (isolated metal species trapped by intrinsic defects) electrocatalytically active centers. Different activated carbon sites have inherent defects or heteroatom doping. Incorporating these metal species is generally considered to change the electronic structure and catalytic properties of adjacent carbon atoms and act as active catalytic centers or participate in construction. Single-atom catalysts (SACs) are highly dispersed metal atoms immobilized on supports, which can maximize the utilization efficiency of metal atoms, thereby achieving high activity, stability, and selectivity of electrocatalytic reactions. SACs with $M-N_x-C$ sites on the carbon matrix of ORR and OER have been extensively studied due to their optimal atom utilization efficiency and high intrinsic activity. Another method to introduce metallic elements into carbon materials is hybrid electrocatalysts, which consist of TM-based particles (including metals, alloys, oxides, nitrides, sulfides, etc.) and carbon supports. In addition, metal–organic frameworks (MOFs) and covalent organic frameworks (COFs) have also been considered as ideal precursors for the rational dispersion of metals in carbon scaffolds.

Assembly of TM compounds (such as oxides, alloys, hydroxides, sulfides, phosphates, nitrides, carbides, peroxides, and spinels) on heteroatom-doped carbon via chemical attachment and electrical coupling, is another feasible method to introduce TM into carbon materials. Fortunately, TM compounds anchored on nanocarbon materials exhibit excellent electrocatalytic activity and persistence in ORR and OER due to their good electrical conductivity, excellent chemical stability, high surface area, robust synergy between TM compounds and heteroatom-doped nanocarbons.

The role of oxygen vacancies in the electrochemical performance of Co_3O_4 has been extensively studied [288–291]. DFT calculations showed that the oxygen vacancy created a new defect state located in the band gap of Co_3O_4, which readily excited two electrons in the defect state, resulting in enhanced conductivity and superior electrochemical activity [292]. S. Wang et al. developed a simple and efficient plasmonic engraving strategy to produce Co_3O_4 nanosheets with oxygen vacancies [56]. It was found that the electrocatalytic performance of Co_3O_4 on OER was mainly affected by its high surface area and oxygen vacancies. This suggests that the introduction of oxygen vacancies can serve as an effective method to further enhance the bifunctional catalytic activity of cobalt oxide catalysts. S. Guo et al. reported that the Kirkendall diffusion process not only preserved the well-designed porous N-doped carbon structure but also induced the generation of oxygen vacancies in Co_3O_4 hollow particles [288]. The oxygen vacancy concentration in the final product can be controlled by simply adjusting the oxidation time to change the surface electronic structure and tune the number of active sites. Benefiting from the tailored oxygen vacancies and unique structural design, the resulting catalysts achieve excellent performance in both ORR and OER. Furthermore, it also shows great promise for portable solid-state ZABs with stable cycling performance, high power density and capacity. Likewise, the introduction of oxygen vacancies in the crystals of ABO_3 peroxides has also been used to enhance the catalytic activity [293–295].

Encapsulation of TM compounds, including alloys, hydroxides, sulfides, phosphates, and nitride nanoparticles, in N-doped carbon is also an efficient approach to generating electrocatalytically active site-rich materials [287, 296–303]. N-doped carbon effectively facilitates electron transport and prevents nanoparticle aggregation. The synergistic effect of TM compounds in contact with carbon materials can enhance the electron transport ability, thereby promoting the catalytic efficiency and stability of ORR and OER. S. Gupta et al. synthesized highly active and stable graphene tubes decorated with FeCoNi alloy nanoparticles [304]. It was found that the unique structure of the graphene tube provided the largest electrochemically available surface area, which, coupled with the optimal degree of graphitization and pyridine nitrogen content, greatly promoted the activity and durability of the catalyst. Furthermore, the ORR and OER activities of FeCoNi-derived N-graphene tubes were superior to those of all other derived N-graphene tubes. M. Wu et al. developed Fe/Co hydroxide/oxide nanoparticles coupled to NCNTs (FeCo-DHO/NCNTs) by direct anchoring, nucleation, and growth [305]. Benefiting from bimetallic synergistic coupling, high electrical contact area, and strong adhesion on conductive carbon supports, the catalyst exhibits high bifunctional ORR/OER activity and excellent charge–discharge performance and long cycle life. X. Sun et al. fabricated a new type of Co_9S_8 nanoparticles/N, S-doped defect-rich carbon nanotubes (Co_9S_8/N, S-CNTs) by optimizing the precursor composition, followed by decomposition and post-treatment under the Ar atmosphere [306]. The intimate contact and synergistic effect between Co_9S_8 and N, S-CNTs resulted in smaller charge transfer resistance, good electronic conductivity, and larger effective electrochemical area, resulting in stronger ORR and OER activity. Furthermore, when Co_9S_8 nanoparticles/N-doped carbon nanotubes (Co_9S_8-NCs) were used as cathode electrocatalysts for flexible solid-state ZABs, at a current density of 5 mA cm^{-2} for 20 min per cycle, it exhibited excellent activity (initial discharge and charge voltages is ~1.10 and 1.94 V) and stability (the voltage gap between charge and discharge decreased slightly from 0.84 V to 0.78 V after 900 min of continuous testing). J. Zhang et al. developed cobalt phosphide (Co_2P)-cobalt nitride (CoN) core–shell nanoparticles (Co_2P/CoNin-NCNTs) with dual active sites through a simple one-step self-assembly and closed pyrolysis method [307]. The topological curvature and abundant defect sites in Co_2P/CoN-in-NCNTs can tune and optimize their electronic structures. Combined with pyrroline N and Co-N active sites, Co_2P/CoN-in-NCNTs can provide efficient bifunctional catalysis and all-solid-state ZABs with good performance. When the flexible all-solid-state battery is subjected to mechanical stress, the voltage remains unchanged, revealing the attractive potential of Co_2P/CoN-in-NCNTs catalysts in rechargeable and flexible all-solid-state ZABs for wearable optoelectronic devices.

Recently, hybrid materials encapsulating transition metals (TMs) in graphitic layers were found to be efficient electrocatalysts for promoting ORR and OER (Fig. 3.37a, b) [308–312]. Transition metal nanoparticles can simultaneously play multiple roles affecting the surface carbon layer, including increasing its degree of graphitization during carbonization and transferring electrons. In turn, the surface carbon layer prevents oxidation, acid leaching, and aggregation of encapsulated transition metal nanoparticles during electrocatalysis [178]. The catalytic activity of these catalysts largely depends on two key factors, including intrinsic activity and the density of active sites. The intrinsic activity is determined by the electronic structure of the surface graphitic layer, which can be tuned by the electronic modulation effect of a suitable metallic core, where the interaction between them changes the local work function of the shell, inducing surprisingly high chemical activity [313]. In addition, doping nitrogen in the carbon lattice can further induce non-uniform charge distribution of adjacent carbon atoms. The additional synergistic effect between doped nitrogen and encapsulated transition metal nanoparticles also stimulates the enhancement of electrocatalytic activity (Fig. 3.37c) [310, 312]. The density of active sites is deeply based on modulating the effective area of surface carbon, which is primarily a function of metal core size, loading, and dispersion.

Generally speaking, the key to this kind of catalyst is relying on high temperature synthesis. J. Wang et al. reported the synthesis of high-density iron nanoparticles encapsulated in nitrogen-doped carbon nanoshells (Fe@N-C) through the rational pyrolysis of metal salts and organic molecular compounds by solid-phase pyrolysis of dicyandiamide and ferric ammonium citrate precursors [178]. During the carbonization process, a high degree of graphitization is generated in the carbon phase with the help of some iron nanoparticles, which can highly promote conductivity during electrocatalysis. Thus, in contrast to the state-of-the-art commercial Pt/C and IrO_2 phases ratio, the obtained Fe@N-C-700 material shows excellent dual functions of ORR and OER in alkaline medium, as well as high Zn-air battery performance. Despite the high performance achieved, the poor compatibility of the physical mixing of organic

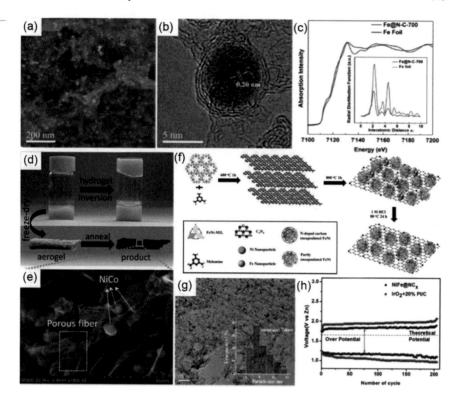

Fig. 3.37 a SEM. **b** Transmission electron microscope (TEM). **c** X-ray absorption near-edge structure (XANES) and extended X-ray absorption fine structure (EXAFS) spectra of Fe@N-C-700 [178]. **d** Fabrication process and **e** scanned image of the 3D Ni-Co/porous fibrous carbon (PFC) aerogel [314]. **f** Schematic diagram of the synthesis strategy and **g** TEM image of the TMs@NCX composite (inset: corresponding particle size distribution histogram). **h** Curves of cycling performed at 10 mA cm^{-2}, each with a duration of 600 s [315]

molecular compounds and metal salts often does not allow for controlled homogeneous distribution of the different components, which inevitably leads to agglomeration of metal particles and microstructural inhomogeneities, thus hindering further improvements. Therefore, better efforts may be needed to adjust the activity in a controlled manner. For example, based on sol–gel chemistry, Fu et al. used a novel $K_2Ni(CN)_4/K_3Co(CN)_6$-chitosan hydrogel system [314] to demonstrate a porous fibrous carbon aerogel and immobilized nickel–cobalt nanoparticles constitute an excellent bifunctional catalyst (nickel–cobalt/porous fibrous carbon (PFC) aerogel) (Fig. 3.37d, e). However, sol–gel methods and subsequent pyrolysis methods often fail to properly control the pore structure.

Given the well-defined coordination environment, ordered arrangement within the framework, and periodic porous structure, MOFs can serve as a perfect encapsulation of metal ions [316–318]; the pyrolysis of MOFs and their compounds is defined as a promising solution to this problem. In addition, tailor-made metal coordination centers and organic ligands, abundant porous structures and high specific surface areas provide them with more possibilities. A variety of different metal and ligand MOF types have been developed, among which iron-based MOFs are the most popular and effective, such as zeolitic imidazole frameworks (ZIFs) [319], iron(III)-based metal–organic frameworks (MILs) [320, 321], and et al. As one of the simplest MOFs [322], the cobalt analog of Prussian blue can be directly used to synthesize bifunctional electrocatalysts of metallic cobalt core/graphene shell. Despite the simplicity of the process, the agglomeration of metal particles is unavoidable due to the implicit structural factor of Prussian blue. Recently, J. Zhu et al. demonstrated catalysts featuring thin graphene nanosheets with fully encapsulated ultrafine and highly loaded ($\approx$25 wt%) transition metal nanoparticles (TMs@NCX) [315], which depend on MOF-based unique two-stage encapsulation strategy for NiFe-MILs and melamine pyrolysis (Fig. 3.37f, g). Based on the regulation of the electronic structure of the outer carbon layers by electron permeation from the NiFe core, the increased active site density of the encapsulated nanoalloys with reduced size, and the enhancement of electron density in the graphene shell, the superior NiFe@NCX catalysts exhibited extremely high performance. It possesses good electrocatalytic activity and high stability, the onset potential of ORR is 1.03 V, and the overpotential of OER at 10 mA cm^{-2} is only 0.23 V. The rechargeable Zn-air battery using NiFe@NCX catalyst exhibited unprecedented long-term cycling stability at 10 mA cm^{-2} (Fig. 3.37h).

References

1. Y. Li, M. Gong, Y. Liang, J. Feng, J.-E. Kim, H. Wang, G. Hong, B. Zhang, H. Dai, Nat. Commun. **2013**, 4 (1805)
2. S. Ravichandran, N. Bhuvanendran, Q. Xu, T. Maiyalagan, L. Xing, H. Su, J. Colloid Interface Sci. **608**, 207 (2022)
3. J.J. Han, N. Li, T.Y. Zhang, J. Power Sources **193**, 885 (2009)
4. T. Wang, M. Kaempgen, P. Nopphawan, G. Wee, S. Mhaisalkar, M. Srinivasan, J. Power Sources **195**, 4350 (2010)
5. M. Chatenet, L. Genies-Bultel, M. Aurousseau, R. Durand, F. Andolfatto, J. Appl. Electrochem. **32**, 1131 (2002)
6. Z. Zhang, K. Yao, L. Cong, Z. Yu, L. Qu, W. Huang, Catal. Sci. Technol. **10**, 1336 (2020)
7. T. Qiu, Z. Liang, W. Guo, S. Gao, C. Qu, H. Tabassum, H. Zhang, B. Zhu, R. Zou, Y. Shao-Horn, Nano Energy **58**, 1 (2019)
8. J. Ying, G. Jiang, Z. Paul Cano, L. Han, X.-Y. Yang, Z. Chen, Nano Energy, **40**, 88 (2017)
9. J. Kim, Y. Lee, S. Sun, J. Am. Chem. Soc. **132**, 4996 (2010)
10. P.P. Fang, S. Duan, X.D. Lin, J.R. Anema, J.F. Li, O. Buriez, Y. Ding, F.R. Fan, D.Y. Wu, B. Ren, Z.L. Wang, C. Amatore, Z.Q. Tian, Chem. Sci. **2**, 531 (2011)
11. S. Zhang, S. Guo, H. Zhu, D. Su, S. Sun, J. Am. Chem. Soc. **134**, 5060 (2012)
12. Y. Jin, F. Chen, Y. Lei, X. Wu, ChemCatChem **7**, 2377 (2015)
13. J. Hu, Q. Liu, L. Shi, Z. Shi, H. Huang, Appl. Surf. Sci. **402**, 61 (2017)
14. C.Y. Chang, C.Y. Chu, Y.C. Huang, C.W. Huang, S.Y. Chang, C.A. Chen, C.Y. Chao, W.F. Su, A.C.S. Appl, Mater. Interfaces **7**, 4955 (2015)
15. V.M. Dhavale, S. Kurungot, ACS Catal. **5**, 1445 (2015)
16. Z. Cui, H. Chen, M. Zhao, F.J. Disalvo, Nano Lett. **16**, 2560 (2016)

17. Y. Jin, F. Chen, Electrochim. Acta **158**, 437 (2015)
18. S. Yuan, Z. Pu, H. Zhou, J. Yu, I.S. Amiinu, J. Zhu, Q. Liang, J. Yang, D. He, Z. Hu, G. Van Tendeloo, S. Mu, Nano Energy **59**, 472 (2019)
19. C. H. Kuo, I. M. Mosa, S. Thanneeru, V. Sharma, L. Zhang, S. Biswas, M. Aindow, S. Pamir Alpay, J. F. Rusling, S. L. Suib, J. He, Chem. Commun., **51**, 5951 (2015)
20. F. Cheng, J. Shen, B. Peng, Y. Pan, Z. Tao, J. Chen, Nat. Chem. **3**, 79 (2011)
21. G. Li, X. Wang, J. Fu, J. Li, M.G. Park, Y. Zhang, G. Lui, Z. Chen, Angew. Chem. Int. Ed. **55**, 4977 (2016)
22. A. Hammouche, A. Kahoul, D.U. Sauer, R.W. De Doncker, J. Power Sources **153**, 239 (2006)
23. J. Tulloch, S.W. Donne, J. Power Sources **188**, 359 (2009)
24. J. Suntivich, H.A. Gasteiger, N. Yabuuchi, H. Nakanishi, J.B. Goodenough, Y. Shao-Horn, Nat. Chem. **3**, 546 (2011)
25. J. Bian, X. Cheng, X. Meng, J. Wang, J. Zhou, S. Li, Y. Zhang, C. Sun, A.C.S. Appl, Energy Mater. **2**, 2296 (2019)
26. G. Cheng, G. Liu, P. Liu, L. Chen, S. Han, J. Han, F. Ye, W. Song, B. Lan, M. Sun, L. Yu, Front. Chem. **7**, 766 (2019)
27. Y. Li, H. Huang, S. Chen, X. Yu, C. Wang, T. Ma, Nano Res. **12**, 2774 (2019)
28. D. Lim, H. Kong, C. Lim, N. Kim, S.E. Shim, S.H. Baeck, Int. J. Hydrogen Energy **44**, 23775 (2019)
29. H. Ge, G. Li, T. Zheng, F. Wang, M. Shao, H. Liu, X. Meng, Electrochim. Acta **319**, 1 (2019)
30. C. Guan, A. Sumboja, H. Wu, W. Ren, X. Liu, H. Zhang, Z. Liu, C. Cheng, S.J. Pennycook, J. Wang, Adv. Mater. **29**, 1704117 (2017)
31. Y. Jiang, Y.-P. Deng, J. Fu, D.U. Lee, R. Liang, Z.P. Cano, Y. Liu, Z. Bai, S. Hwang, L. Yang, D. Su, W. Chu, Z. Chen, Adv. Energy Mater. **8**, 1702900 (2018)
32. X.F. Lu, Y. Chen, S. Wang, S. Gao, X.W. Lou, Adv. Mater. **31**, 1902339 (2019)
33. V. Neburchilov, H. Wang, J.J. Martin, W. Qu, J. Power Sources **195**, 1271 (2010)
34. J. Xu, C. Chen, Z. Han, Y. Yang, J. Li, Q. Deng, Nanomaterials **9**, 1161 (2019)
35. J. Horkans, M.W. Shafer, J. Electrochem. Soc. **124**, 1202 (1977)
36. K.E. Sickafus, J.M. Wills, N.W. Grimes, J. Am. Ceram. Soc. **82**, 3279 (1999)
37. R.J. Hill, J.R. Craig, G.V. Gibbs, Phys. Chem. Miner. **4**, 317 (1979)
38. Q. Zhao, Z. Yan, C. Chen, J. Chen, Chem. Rev. **117**, 10121 (2017)
39. C. Li, X. Han, F. Cheng, Y. Hu, C. Chen, J. Chen, Nat. Commun. **6**, 7345 (2015)
40. G. Wu, J. Wang, W. Ding, Y. Nie, L. Li, X. Qi, S. Chen, Z. Wei, Angew. Chem. Int. Ed. **55**, 1340 (2016)
41. X. Ge, Y. Liu, F.W.T. Goh, T.S.A. Hor, Y. Zong, P. Xiao, Z. Zhang, S.H. Lim, B. Li, X. Wang, Z. Liu, A.C.S. Appl, Mater. Interfaces **6**, 12684 (2014)
42. X. Han, G. He, Y. He, J. Zhang, X. Zheng, L. Li, C. Zhong, W. Hu, Y. Deng, T.Y. Ma, Adv. Energy Mater. **8**, 1702222 (2018)
43. Q. Liu, Z. Chen, Z. Yan, Y. Wang, E. Wang, S. Wang, S. Wang, G. Sun, ChemElectroChem **5**, 1080 (2018)
44. T. Maiyalagan, K.A. Jarvis, S. Therese, P.J. Ferreira, A. Manthiram, Nat. Commun. **5**, 3949 (2014)
45. Z. Du, P. Yang, L. Wang, Y. Lu, J.B. Goodenough, J. Zhang, D. Zhang, J. Power Sources **265**, 91 (2014)
46. H. Sun, Y. Chen, C. Xu, D. Zhu, L. Huang, J. Solid State Electrochem. **16**, 1247 (2012)
47. L. Xu, Z. Wang, J. Wang, Z. Xiao, X. Huang, Z. Liu, S. Wang, Nanotechnology **28**, 165402 (2017)
48. S. Niu, W.J. Jiang, Z. Wei, T. Tang, J. Ma, J.S. Hu, L.J. Wan, J. Am. Chem. Soc. **141**, 7005 (2019)
49. J. Liu, H. Liu, F. Wang, Y. Song, RSC Adv. **5**, 90785 (2015)
50. P. Sivakumar, P. Subramanian, T. Maiyalagan, A. Gedanken, A. Schechter, Mater. Chem. Phys. **229**, 190 (2019)
51. K. Chakrapani, G. Bendt, H. Hajiyani, T. Lunkenbein, M.T. Greiner, L. Masliuk, S. Salamon, J. Landers, R. Schlögl, H. Wende, R. Pentcheva, S. Schulz, M. Behrens, ACS Catal. **8**, 1259 (2018)

52. L. Huang, D. Chen, G. Luo, Y.R. Lu, C. Chen, Y. Zou, C.L. Dong, Y. Li, S. Wang, Adv. Mater. **31**, 1901439 (2019)
53. J. Sun, N. Guo, Z. Shao, K. Huang, Y. Li, F. He, Q. Wang, Adv. Energy Mater. **8**, 1800980 (2018)
54. S. Peng, F. Gong, L. Li, D. Yu, D. Ji, T. Zhang, Z. Hu, Z. Zhang, S. Chou, Y. Du, S. Ramakrishna, J. Am. Chem. Soc. **140**, 13644 (2018)
55. Q. Kang-Wen, C. Xi, Y. Zhang, R. Zhang, Z. Li, G.-R. Sheng, H. Liu, C.-K. Dong, Y.-J. Chen, X.-W. Du, Chem. Commun. **55**, 8579 (2019)
56. L. Xu, Q. Jiang, Z. Xiao, X. Li, J. Huo, S. Wang, L. Dai, Angew. Chem. Int. Ed. **55**, 5277 (2016)
57. G. Zhang, J. Yang, H. Wang, H. Chen, J. Yang, F. Pan, A.C.S. Appl, Mater. Interfaces **9**, 16159 (2017)
58. Z.Q. Liu, H. Cheng, N. Li, T.Y. Ma, Y.Z. Su, Adv. Mater. **28**, 3777 (2016)
59. M. Fayette, A. Nelson, R.D. Robinson, J. Mater. Chem. A **3**, 4274 (2015)
60. S.K. Singh, V.M. Dhavale, S. Kurungot, A.C.S. Appl, Mater. Interfaces **7**, 21138 (2015)
61. J. Xu, P. Gao, T.S. Zhao, Energy Environ. Sci. **5**, 5333 (2012)
62. S.G. Mohamed, Y.Q. Tsai, C.J. Chen, Y.T. Tsai, T.F. Hung, W.S. Chang, R.S. Liu, A.C.S. Appl, Mater. Interfaces **7**, 12038 (2015)
63. J.T. Ren, G.G. Yuan, C.C. Weng, Z.Y. Yuan, A.C.S. Sustain, Chem. Eng. **6**, 707 (2018)
64. J.G. Kim, Y. Kim, Y. Noh, W.B. Kim, Chemsuschem **8**, 1752 (2015)
65. H. Wang, S. Zhuo, Y. Liang, X. Han, B. Zhang, Angew. Chem. Int. Ed. **55**, 9055 (2016)
66. G. Y. Zhang, B. Guo, J. Chen, Sens. Actuators, B, **114**, 402 (2006)
67. M. Li, Y. Xiong, X. Liu, X. Bo, Y. Zhang, C. Han, L. Guo, Nanoscale **7**, 8920 (2015)
68. S.M. Hwang, S.Y. Kim, J.G. Kim, K.J. Kim, J.W. Lee, M.S. Park, Y.J. Kim, M. Shahabuddin, Y. Yamauchi, J.H. Kim, Nanoscale **7**, 8351 (2015)
69. X. Tong, S. Chen, C. Guo, X. Xia, X.Y. Guo, A.C.S. Appl, Mater. Interfaces **8**, 28274 (2016)
70. T. Odedairo, X. Yan, J. Ma, Y. Jiao, X. Yao, A. Du, Z. Zhu, A.C.S. Appl, Mater. Interfaces **7**, 21373 (2015)
71. X. Li, L. Yuan, J. Wang, L. Jiang, A.I. Rykov, D.L. Nagy, C. Bogdán, M.A. Ahmed, K. Zhu, G. Sun, W. Yang, Nanoscale **8**, 2333 (2016)
72. S.K. Singh, V.M. Dhavale, S. Kurungot, A.C.S. Appl, Mater. Interfaces **7**, 442 (2015)
73. J. Feng, H.C. Zeng, Chem. Mater. **15**, 2829 (2003)
74. L. Leng, X. Zeng, H. Song, T. Shu, H. Wang, S. Liao, J. Mater. Chem. A **3**, 15626 (2015)
75. S. Lin, X. Shi, H. Yang, D. Fan, Y. Wang, K. Bi, J. Alloys Compd. **720**, 147 (2017)
76. H. Wang, R. Liu, Y. Li, X. Lü, Q. Wang, S. Zhao, K. Yuan, Z. Cui, X. Li, S. Xin, R. Zhang, M. Lei, Z. Lin, Joule **2**, 337 (2018)
77. J. Li, S. Xiong, X. Li, Y. Qian, Nanoscale **5**, 2045 (2013)
78. Y. Li, P. Hasin, Y. Wu, Adv. Mater. **2010**, 22 (1926)
79. B. Lu, D. Cao, P. Wang, G. Wang, Y. Gao, Int. J. Hydrogen Energy **36**, 72 (2011)
80. J. Shi, K. Lei, W. Sun, F. Li, F. Cheng, J. Chen, Nano Res. **10**, 3836 (2017)
81. H. Cheng, M.-L. Li, C.-Y. Su, N. Li, Z.-Q. Liu, Adv. Funct. Mater. **27**, 1701833 (2017)
82. H. Yang, Y. Liu, S. Luo, Z. Zhao, X. Wang, Y. Luo, Z. Wang, J. Jin, J. Ma, ACS Catal. **7**, 5557 (2017)
83. J. Du, C. Chen, F. Cheng, J. Chen, Inorg. Chem. **54**, 5467 (2015)
84. K.-N. Jung, S.M. Hwang, M.-S. Park, K.J. Kim, J.-G. Kim, S.X. Dou, J.H. Kim, J.-W. Lee, Sci. Rep. **5**, 7665 (2015)
85. X. Wang, Y. Li, T. Jin, J. Meng, L. Jiao, M. Zhu, J. Chen, Nano Lett. **17**, 7989 (2017)
86. D.U. Lee, M.G. Park, Z.P. Cano, W. Ahn, Z. Chen, Chemsuschem **11**, 406 (2018)
87. C. Jin, F. Lu, X. Cao, Z. Yang, R. Yang, J. Mater. Chem. A **1**, 12170 (2013)
88. J.G. Kim, Y. Noh, Y. Kim, S. Lee, W.B. Kim, Nanoscale **9**, 5119 (2017)
89. P. Li, W. Sun, Q. Yu, P. Yang, J. Qiao, Z. Wang, D. Rooney, K. Sun, Solid State Ionics **289**, 17 (2016)
90. X. Cao, W. Yan, C. Jin, J. Tian, K. Ke, R. Yang, Electrochim. Acta **180**, 788 (2015)
91. C. Shenghai, S. Liping, K. Fanhao, H. Lihua, Z. Hui, J. Power Sources **430**, 25 (2019)

92. J. Han, S. Hao, Z. Liu, A.M. Asiri, X. Sun, Y. Xu, Chem. Commun. **54**, 1077 (2018)
93. X. Cao, Z. Sun, X. Zheng, J. Tian, C. Jin, R. Yang, F. Li, P. He, H. Zhou, J. Mater. Chem. A **5**, 19991 (2017)
94. X. Cao, C. Jin, F. Lu, Z. Yang, M. Shen, R. Yang, J. Electrochem. Soc. **161**, H296 (2014)
95. J. Zhao, Y. He, Z. Chen, X. Zheng, X. Han, D. Rao, C. Zhong, W. Hu, Y. Deng, A.C.S. Appl, Mater. Interfaces **11**, 4915 (2019)
96. M.A. Peña, J.L.G. Fierro, Chem. Rev. **2001**, 101 (1981)
97. H. Wang, M. Zhou, P. Choudhury, H. Luo, Appl. Mater. Today **16**, 56 (2019)
98. A. Chilvery, S. Das, P. Guggilla, C. Brantley, A. Sunda-Meya, Sci. Technol. Adv. Mater. **17**, 650 (2016)
99. T. Hyodo, M. Hayashi, N. Miura, N. Yamazoe, J. Electrochem. Soc. **143**, L266 (1996)
100. C. Zhu, A. Nobuta, I. Nakatsugawa, T. Akiyama, Int. J. Hydrogen Energy **38**, 13238 (2013)
101. Y. Zhao, Y. Hang, Y. Zhang, Z. Wang, Y. Yao, X. He, C. Zhang, D. Zhang, Electrochim. Acta **232**, 296 (2017)
102. Z. Wu, L.P. Sun, T. Xia, L.H. Huo, H. Zhao, A. Rougier, J.C. Grenier, J. Power Sources **334**, 86 (2016)
103. S. Velraj, J.H. Zhu, J. Power Sources **227**, 48 (2013)
104. Y. Shimizu, H. Matsuda, N. Miura, N. Yamazoe, Chem. Lett. **21**, 1033 (1992)
105. Z. Wang, C. Jin, J. Sui, C. Li, R. Yang, Int. J. Hydrogen Energy **43**, 20727 (2018)
106. S. Peng, X. Han, L. Li, S. Chou, D. Ji, H. Huang, Y. Du, J. Liu, S. Ramakrishna, Adv. Energy Mater. **8**, 1800612 (2018)
107. Z. Zhang, Y. Zhu, Y. Zhong, W. Zhou, Z. Shao, Adv. Energy Mater. **7**, 1700242 (2017)
108. Y. Zhu, W. Zhou, J. Sunarso, Y. Zhong, Z. Shao, Adv. Funct. Mater. **26**, 5862 (2016)
109. Z. Li, L. Lv, J. Wang, X. Ao, Y. Ruan, D. Zha, G. Hong, Q. Wu, Y. Lan, C. Wang, J. Jiang, M. Liu, Nano Energy **47**, 199 (2018)
110. J. Wang, Y. Gao, D. Chen, J. Liu, Z. Zhang, Z. Shao, F. Ciucci, ACS Catal. **8**, 364 (2018)
111. J. Suntivich, K.J. May, H.A. Gasteiger, J.B. Goodenough, Y. Shao-Horn, Science **334**, 1383 (2011)
112. Y. Luo, F. Lu, C. Jin, Y. Wang, R. Yang, C. Yang, J. Power Sources **319**, 19 (2016)
113. S. Bie, Y. Zhu, J. Su, C. Jin, S. Liu, R. Yang, J. Wu, J. Mater. Chem. A **3**, 22448 (2015)
114. J.J. Xu, Z.L. Wang, D. Xu, F.Z. Meng, X.B. Zhang, Energy Environ. Sci. **7**, 2213 (2014)
115. J.J. Xu, D. Xu, Z.L. Wang, H.G. Wang, L.L. Zhang, X.B. Zhang, Angew. Chem. Int. Ed. **52**, 3887 (2013)
116. Y. Bu, O. Gwon, G. Nam, H. Jang, S. Kim, Q. Zhong, J. Cho, G. Kim, ACS Nano **11**, 11594 (2017)
117. N.I. Kim, S.H. Cho, S.H. Park, Y.J. Lee, R.A. Afzal, J. Yoo, Y.S. Seo, Y.J. Lee, J.Y. Park, J. Mater. Chem. A **6**, 17807 (2018)
118. T. Lippert, M.J. Montenegro, M. Döbeli, A. Weidenkaff, S. Müller, P.R. Willmott, A. Wokaun, Prog. Solid State Chem. **35**, 221 (2007)
119. K. A. Stoerzinger, M. Risch, J. Suntivich, W. M. Lü, J. Zhou, M. D. Biegalski, H. M. Christen, Ariando, T. Venkatesan, S. H. Yang, *Energy Environ. Sci.*, **6**, 1582 (2013)
120. E. Fabbri, R. Mohamed, P. Levecque, O. Conrad, R. Kötz, T.J. Schmidt, ChemElectroChem **1**, 338 (2014)
121. F. Yusoff, N. Mohamed, A. Aziz, S.A. Ghani, Mater. Sci. Appl. **05**, 199 (2014)
122. E. Fabbri, R. Mohamed, P. Levecque, O. Conrad, R. Kötz, T.J. Schmidt, ACS Catal. **4**, 1061 (2014)
123. C. Jin, X. Cao, F. Lu, Z. Yang, R. Yang, Int. J. Hydrogen Energy **38**, 10389 (2013)
124. J. Kim, X. Chen, P.C. Shih, H. Yang, A.C.S. Sustain, Chem. Eng. **5**, 10910 (2017)
125. A. Vignesh, M. Prabu, S. Shanmugam, A.C.S. Appl, Mater. Interfaces **8**, 6019 (2016)
126. X. Chu, K. Huang, M. Han, S. Feng, Inorg. Chem. **52**, 4130 (2013)
127. G. Liu, H. Chen, L. Xia, S. Wang, L.X. Ding, D. Li, K. Xiao, S. Dai, H. Wang, A.C.S. Appl, Mater. Interfaces **7**, 22478 (2015)
128. K.N. Jung, J.H. Jung, W.B. Im, S. Yoon, K.H. Shin, J.W. Lee, A.C.S. Appl, Mater. Interfaces **5**, 9902 (2013)

129. Y. Wang, Z. Yang, F. Lu, C. Jin, J. Wu, M. Shen, R. Yang, F. Chen, RSC Adv. **5**, 974 (2015)
130. B. Zhao, L. Zhang, D. Zhen, S. Yoo, Y. Ding, D. Chen, Y. Chen, Q. Zhang, B. Doyle, X. Xiong, M. Liu, Nat. Commun. **8**, 14586 (2017)
131. F. Azizi, A. Kahoul, A. Azizi, J. Alloys Compd. **484**, 555 (2009)
132. M. Cheriti, A. Kahoul, A. Azizi, N. Alonso-Vante, Ionics **19**, 1155 (2013)
133. C. Jin, Z. Yang, X. Cao, F. Lu, R. Yang, Int. J. Hydrogen Energy **39**, 2526 (2014)
134. X. Han, Y. Hu, J. Yang, F. Cheng, J. Chen, Chem. Commun. **50**, 1497 (2014)
135. Z. Wang, F. Zhang, C. Jin, Y. Luo, J. Sui, H. Gong, R. Yang, Carbon **115**, 261 (2017)
136. Y. Wang, Z. Wang, C. Jin, C. Li, X. Li, Y. Li, R. Yang, M. Liu, Electrochim. Acta **318**, 120 (2019)
137. Y.Q. Zhang, H.B. Tao, Z. Chen, M. Li, Y.F. Sun, B. Hua, J.L. Luo, J. Mater. Chem. A **7**, 26607 (2019)
138. R. S. Kalubarme, M. S. Cho, J. K. Kim, C. J. Park, *Nanotechnology*, **23** (2012)
139. T. Masuda, H. Fukumitsu, K. Fugane, H. Togasaki, D. Matsumura, K. Tamura, Y. Nishihata, H. Yoshikawa, K. Kobayashi, T. Mori, K. Uosaki, J. Phys. Chem. C **116**, 10098 (2012)
140. K. Fugane, T. Mori, D.R. Ou, A. Suzuki, H. Yoshikawa, T. Masuda, K. Uosaki, Y. Yamashita, S. Ueda, K. Kobayashi, N. Okazaki, I. Matolinova, V. Matolin, Electrochim. Acta **56**, 3874 (2011)
141. Y.C. Wei, C.W. Liu, K.W. Wang, ChemPhysChem **11**, 3078 (2010)
142. C. Jin, X. Cao, F. Lu, Z. Yang, R. Yang, A.C.S. Appl, Mater. Interfaces **6**, 847 (2014)
143. Q. He, S. Mukerjee, R. Zeis, S. Parres-Esclapez, M. J. Illán-Gómez, A. Bueno-López, Appl. Catal., A, **381**, 54 (2010)
144. C.H. Ahn, R.S. Kalubarme, Y.H. Kim, K.N. Jung, K.H. Shin, C.J. Park, Electrochim. Acta **117**, 18 (2014)
145. R.S. Kalubarme, H.S. Jadhav, C.N. Park, K.N. Jung, K.H. Shin, C.J. Park, J. Mater. Chem. A **2**, 13024 (2014)
146. Y. Zhu, S. Liu, C. Jin, S. Bie, R. Yang, J. Wu, J. Mater. Chem. A **3**, 13563 (2015)
147. F. Song, L. Bai, A. Moysiadou, S. Lee, C. Hu, L. Liardet, X. Hu, J. Am. Chem. Soc. **140**, 7748 (2018)
148. F. Cheng, Y. Su, J. Liang, Z. Tao, J. Chen, Chem. Mater. **22**, 898 (2010)
149. L. Mao, D. Zhang, T. Sotomura, K. Nakatsu, N. Koshiba, T. Ohsaka, Electrochim. Acta **48**, 1015 (2003)
150. J.S. Lee, G.S. Park, H.I. Lee, S.T. Kim, R. Cao, M. Liu, J. Cho, Nano Lett. **11**, 5362 (2011)
151. L. Trotochaud, J.K. Ranney, K.N. Williams, S.W. Boettcher, J. Am. Chem. Soc. **134**, 17253 (2012)
152. X. Guo, T. Wang, P. Liu, H. Zhang, D. Guo, Appl. Catal., B, **250**, 71 (2019)
153. W. Wang, L. Kuai, W. Cao, M. Huttula, S. Ollikkala, T. Ahopelto, A.P. Honkanen, S. Huotari, M. Yu, B. Geng, Angew. Chem. Int. Ed. **56**, 14977 (2017)
154. B. Li, C. Zhao, S. Chen, J. Liu, X. Chen, L. Song, Q. Zhang, Adv. Mater. **31**, 1900592 (2019)
155. X. Zou, A. Goswami, T. Asefa, J. Am. Chem. Soc. **135**, 17242 (2013)
156. T. Huang, T. Fang, L. Xu, Z. Wang, Y. Xin, J. Yu, S. Yao, Z. Zhang, J. Catal. **359**, 223 (2018)
157. L. Zhu, D. Susac, M. Teo, K.C. Wong, P.C. Wong, R.R. Parsons, D. Bizzotto, K.A.R. Mitchell, S.A. Campbell, J. Catal. **258**, 235 (2008)
158. S. Shit, S. Chhetri, W. Jang, N.C. Murmu, H. Koo, P. Samanta, T. Kuila, A.C.S. Appl, Mater. Interfaces **10**, 27712 (2018)
159. X. Guo, Y. Qian, W. Zhang, C. Qian, F. Xu, S. Qian, H. Yang, A. Yuan, T. Fan, J. Alloys Compd. **765**, 835 (2018)
160. C. Liu, F. Dong, M. Wu, Y. Wang, N. Xu, X. Wang, J. Qiao, P. Shi, H. Huang, J. Power Sources **438**, 226953 (2019)
161. H. Yang, B. Wang, H. Li, B. Ni, K. Wang, Q. Zhang, X. Wang, Adv. Energy Mater. **8**, 1801839 (2018)
162. T. Meng, J. Qin, S. Wang, D. Zhao, B. Mao, M. Cao, J. Mater. Chem. A **5**, 7001 (2017)
163. S.H. Ahn, A. Manthiram, Small **13**, 1702068 (2017)
164. Z. Zeng, L. Yi, J. He, Q. Hu, Y. Liao, Y. Wang, W. Luo, M. Pan, J. Mater. Sci. **55**, 4780 (2020)

165. J. Han, X. Meng, L. Lu, J. Bian, Z. Li, C. Sun, Adv. Funct. Mater. **29**, 1808872 (2019)
166. S. Li, L. Zhang, S. Jie, Z. Liu, New J. Chem. **44**, 5404 (2020)
167. I.S. Amiinu, X. Liu, Z. Pu, W. Li, Q. Li, J. Zhang, H. Tang, H. Zhang, S. Mu, Adv. Funct. Mater. **28**, 1704638 (2018)
168. C. Li, E. Zhou, Z. Yu, H. Liu and M. Xiong, Appl. Catal., B, **269**, 118771 (2020)
169. J. Wu, H. Zhou, Q. Li, M. Chen, J. Wan, N. Zhang, L. Xiong, S. Li, B.Y. Xia, G. Feng, M. Liu, L. Huang, Adv. Energy Mater. **9**, 1900149 (2019)
170. D. Ji, L. Fan, L. Li, S. Peng, D. Yu, J. Song, S. Ramakrishna, S. Guo, Adv. Mater. **31**, 1808267 (2019)
171. L. Ma, S. Chen, Z. Pei, Y. Huang, G. Liang, F. Mo, Q. Yang, J. Su, Y. Gao, J.A. Zapien, C. Zhi, ACS Nano **2018**, 12 (1949)
172. M. Zhang, Q. Dai, H. Zheng, M. Chen, L. Dai, Adv. Mater. **30**, 1705431 (2018)
173. J. Zhao, H. Hu, M. Wu, Nanoscale **12**, 3750 (2020)
174. H.-F. Wang, C. Tang, Q. Zhang, Adv. Funct. Mater. **28**, 1803329 (2018)
175. Y. Guan, Y. Li, S. Luo, X. Ren, L. Deng, L. Sun, H. Mi, P. Zhang, J. Liu, Appl. Catal., B, **256**, 117871 (2019)
176. X.-F. Yang, A. Wang, B. Qiao, J. Li, J. Liu, T. Zhang, Acc. Chem. Res. **46**, 1740 (2013)
177. Y. Guo, P. Yuan, J. Zhang, Y. Hu, I.S. Amiinu, X. Wang, J. Zhou, H. Xia, Z. Song, Q. Xu, S. Mu, ACS Nano **2018**, 12 (1894)
178. J. Wang, H. Wu, D. Gao, S. Miao, G. Wang, X. Bao, Nano Energy **13**, 387 (2015)
179. F. Meng, H. Zhong, J. Yan, X. Zhang, Nano Res. **10**, 4436 (2017)
180. C.Y. Su, H. Cheng, W. Li, Z.Q. Liu, N. Li, Z. Hou, F.Q. Bai, H.X. Zhang, T.Y. Ma, Adv. Energy Mater. **7**, 1602420 (2017)
181. H. Wu, H. Li, X. Zhao, Q. Liu, J. Wang, J. Xiao, S. Xie, R. Si, F. Yang, S. Miao, X. Guo, G. Wang, X. Bao, Energy Environ. Sci. **9**, 3736 (2016)
182. C. Tang, B. Wang, H.F. Wang, Q. Zhang, Adv. Mater. **29**, 1703185 (2017)
183. S. Li, C. Cheng, X. Zhao, J. Schmidt, A. Thomas, Angew. Chem. Int. Ed. **2018**, 57 (1856)
184. B. Chen, X. He, F. Yin, H. Wang, D.-J. Liu, R. Shi, J. Chen, H. Yin, Adv. Funct. Mater. **27**, 1700795 (2017)
185. T. Wang, Z. Kou, S. Mu, J. Liu, D. He, I.S. Amiinu, W. Meng, K. Zhou, Z. Luo, S. Chaemchuen, F. Verpoort, Adv. Funct. Mater. **28**, 1705048 (2018)
186. S. Wang, J. Qin, T. Meng, M. Cao, Nano Energy **39**, 626 (2017)
187. R. Wang, J. Cao, S. Cai, X. Yan, J. Li, W.M. Yourey, W. Tong, H. Tang, A.C.S. Appl, Energy Mater. **1**, 1060 (2018)
188. S. Liu, M. Wang, X. Sun, N. Xu, J. Liu, Y. Wang, T. Qian, C. Yan, Adv. Mater. **30**, 1704898 (2018)
189. B. Li, Y. Chen, X. Ge, J. Chai, X. Zhang, T.S.A. Hor, G. Du, Z. Liu, H. Zhang, Y. Zong, Nanoscale **8**, 5067 (2016)
190. J. Fu, Z.P. Cano, M.G. Park, A. Yu, M. Fowler, Z. Chen, Adv. Mater. **29**, 1604685 (2017)
191. T. Zhang, Z. Tao, J. Chen, Mater. Horiz. **1**, 196 (2014)
192. J.S. Lee, S.T. Kim, R. Cao, N.S. Choi, M. Liu, K.T. Lee, J. Cho, Adv. Energy Mater. **1**, 34 (2011)
193. Y. Li, J. Lu, ACS Energy Lett. **2**, 1370 (2017)
194. R.D. McKerracher, C. Poncedeleon, R.G.A. Wills, A.A. Shah, F.C. Walsh, ChemPlusChem **80**, 323 (2015)
195. Y. Li, H. Dai, Chem. Soc. Rev. **43**, 5257 (2014)
196. J. Pan, Y.Y. Xu, H. Yang, Z. Dong, H. Liu, B.Y. Xia, Adv. Sci. **5**, 1700691 (2018)
197. W. Zhang, Y. Liu, L. Zhang, J. Chen, Nanomaterials, **9** (2019)
198. A.R. Mainar, E. Iruin, L.C. Colmenares, A. Kvasha, I. de Meatza, M. Bengoechea, O. Leonet, I. Boyano, Z. Zhang, J.A. Blazquez, J. Energy Storage **15**, 304 (2018)
199. Y. Pan, S. Liu, K. Sun, X. Chen, B. Wang, K. Wu, X. Cao, W.C. Cheong, R. Shen, A. Han, Z. Chen, L. Zheng, J. Luo, Y. Lin, Y. Liu, D. Wang, Q. Peng, Q. Zhang, C. Chen, Y. Li, Angew. Chem. Int. Ed. **57**, 8614 (2018)
200. H. Yu, A. Fisher, D. Cheng, D. Cao, A.C.S. Appl, Mater. Interfaces **8**, 21431 (2016)

201. X.X. Wang, D.A. Cullen, Y.T. Pan, S. Hwang, M. Wang, Z. Feng, J. Wang, M.H. Engelhard, H. Zhang, Y. He, Y. Shao, D. Su, K.L. More, J.S. Spendelow, G. Wu, Adv. Mater. **30**, 1706758 (2018)
202. X. Han, X. Ling, D. Yu, D. Xie, L. Li, S. Peng, C. Zhong, N. Zhao, Y. Deng, W. Hu, Adv. Mater. **31**, 1905622 (2019)
203. Z. Lu, B. Wang, Y. Hu, W. Liu, Y. Zhao, R. Yang, Z. Li, J. Luo, B. Chi, Z. Jiang, M. Li, S. Mu, S. Liao, J. Zhang, X. Sun, Angew. Chem. Int. Ed. **58**, 2622 (2019)
204. Z. Yang, Y. Wang, M. Zhu, Z. Li, W. Chen, W. Wei, T. Yuan, Y. Qu, Q. Xu, C. Zhao, X. Wang, P. Li, Y. Li, Y. Wu, Y. Li, ACS Catal. **9**, 2158 (2019)
205. Y. Chen, S. Ji, S. Zhao, W. Chen, J. Dong, W.-C. Cheong, R. Shen, X. Wen, L. Zheng, A.I. Rykov, S. Cai, H. Tang, Z. Zhuang, C. Chen, Q. Peng, D. Wang, Y. Li, Nat. Commun. **9**, 5422 (2018)
206. W. Wei, X. Shi, P. Gao, S. Wang, W. Hu, X. Zhao, Y. Ni, X. Xu, Y. Xu, W. Yan, H. Ji, M. Cao, Nano Energy **52**, 29 (2018)
207. J. Zhang, M. Zhang, Y. Zeng, J. Chen, L. Qiu, H. Zhou, C. Sun, Y. Yu, C. Zhu, Z. Zhu, Small **15**, 1900307 (2019)
208. D. Liu, S. Ding, C. Wu, W. Gan, C. Wang, D. Cao, Z.U. Rehman, Y. Sang, S. Chen, X. Zheng, Y. Wang, B. Ge, L. Song, J. Mater. Chem. A **6**, 6840 (2018)
209. J. Wang, W. Liu, G. Luo, Z. Li, C. Zhao, H. Zhang, M. Zhu, Q. Xu, X. Wang, C. Zhao, Y. Qu, Z. Yang, T. Yao, Y. Li, Y. Lin, Y. Wu, Y. Li, Energy Environ. Sci. **11**, 3375 (2018)
210. W. Zang, A. Sumboja, Y. Ma, H. Zhang, Y. Wu, S. Wu, H. Wu, Z. Liu, C. Guan, J. Wang, S.J. Pennycook, ACS Catal. **8**, 8961 (2018)
211. S. Chen, S. Chen, B. Zhang, J. Zhang, A.C.S. Appl, Mater. Interfaces **11**, 16720 (2019)
212. L. Yang, L. Shi, D. Wang, Y. Lv, D. Cao, Nano Energy **50**, 691 (2018)
213. H.J. Qiu, P. Du, K. Hu, J. Gao, H. Li, P. Liu, T. Ina, K. Ohara, Y. Ito, M. Chen, Adv. Mater. **31**, 1900843 (2019)
214. Y. Yang, C. Wang, S. Gao, K. Mao, G. Xia, Z. Lin, P. Jiang, L. Hu, Q. Chen, Nanoscale **10**, 21076 (2018)
215. C. Du, X. Gao, W. Chen, Chin. J. Catal. **37**, 1049 (2016)
216. J. Li, S. Chen, N. Yang, M. Deng, S. Ibraheem, J. Deng, J. Li, L. Li, Z. Wei, Angew. Chem. Int. Ed. **58**, 7035 (2019)
217. S. Ni, H. Zhang, Y. Zhao, X. Li, Y. Sun, J. Qian, Q. Xu, P. Gao, D. Wu, K. Kato, M. Yamauchi, Y. Sun, Chem. Eng. J. **366**, 631 (2019)
218. W. Xiang, Y. Zhao, Z. Jiang, X. Li, H. Zhang, Y. Sun, Z. Ning, F. Du, P. Gao, J. Qian, K. Kato, M. Yamauchi, Y. Sun, J. Mater. Chem. A **6**, 23366 (2018)
219. J. Li, H. Liu, M. Wang, C. Lin, W. Yang, J. Meng, Y. Xu, K.A. Owusu, B. Jiang, C. Chen, D. Fan, L. Zhou, L. Mai, Chem. Commun. **55**, 334 (2019)
220. K. Yuan, D. Lützenkirchen-Hecht, L. Li, L. Shuai, Y. Li, R. Cao, M. Qiu, X. Zhuang, M.K.H. Leung, Y. Chen, U. Scherf, J. Am. Chem. Soc. **142**, 2404 (2020)
221. Z. Yang, B. Chen, W. Chen, Y. Qu, F. Zhou, C. Zhao, Q. Xu, Q. Zhang, X. Duan, Y. Wu, Nat. Commun. **10**, 3734 (2019)
222. N. K. Wagh, S. S. Shinde, C. H. Lee, J.-Y. Jung, D.-H. Kim, S.-H. Kim, C. Lin, S. U. Lee, J.-H. Lee, Appl. Catal., B, **268**, 118746 (2020)
223. Z. Kou, W. Zang, Y. Ma, Z. Pan, S. Mu, X. Gao, B. Tang, M. Xiong, X. Zhao, A.K. Cheetham, L. Zheng, J. Wang, Nano Energy **67**, 104288 (2020)
224. A.M. Patel, S. Ringe, S. Siahrostami, M. Bajdich, J.K. Nørskov, A.R. Kulkarni, J. Phys. Chem. C **122**, 29307 (2018)
225. L. Chen, Y. Zhang, L. Dong, W. Yang, X. Liu, L. Long, C. Liu, S. Dong, J. Jia, J. Mater. Chem. A **8**, 4369 (2020)
226. J. Chen, H. Li, C. Fan, Q. Meng, Y. Tang, X. Qiu, G. Fu, T. Ma, Adv. Mater. **32**, 2003134 (2020)
227. Z. Li, W. Niu, Z. Yang, N. Zaman, W. Samarakoon, M. Wang, A. Kara, M. Lucero, M.V. Vyas, H. Cao, H. Zhou, G.E. Sterbinsky, Z. Feng, Y. Du, Y. Yang, Energy Environ. Sci. **13**, 884 (2020)

228. C. Xia, Y. Qiu, Y. Xia, P. Zhu, G. King, X. Zhang, Z. Wu, J.Y. Kim, D.A. Cullen, D. Zheng, P. Li, M. Shakouri, E. Heredia, P. Cui, H.N. Alshareef, Y. Hu, H. Wang, Nat. Chem. **13**, 887 (2021)
229. J. Li, M. Chen, D.A. Cullen, S. Hwang, M. Wang, B. Li, K. Liu, S. Karakalos, M. Lucero, H. Zhang, C. Lei, H. Xu, G.E. Sterbinsky, Z. Feng, D. Su, K.L. More, G. Wang, Z. Wang, G. Wu, Nat. Catal. **1**, 935 (2018)
230. Y. Jiang, Y.-P. Deng, R. Liang, J. Fu, R. Gao, D. Luo, Z. Bai, Y. Hu, A. Yu, Z. Chen, Nat. Commun. **11**, 5858 (2020)
231. Y. Lian, W. Yang, C. Zhang, H. Sun, Z. Deng, W. Xu, L. Song, Z. Ouyang, Z. Wang, J. Guo, Y. Peng, Angew. Chem. Int. Ed. **59**, 286 (2020)
232. P. Peng, L. Shi, F. Huo, C. Mi, X. Wu, S. Zhang, Z. Xiang, Sci. Adv., **5**, eaaw2322 (2019)
233. H. Zhong, K.H. Ly, M. Wang, Y. Krupskaya, X. Han, J. Zhang, J. Zhang, V. Kataev, B. Büchner, I.M. Weidinger, S. Kaskel, P. Liu, M. Chen, R. Dong, X. Feng, Angew. Chem. **131**, 10787 (2019)
234. K. Wang, J. Liu, Z. Tang, L. Li, Z. Wang, M. Zubair, F. Ciucci, L. Thomsen, J. Wright, N.M. Bedford, J. Mater. Chem. A **9**, 13044 (2021)
235. M. Tong, F. Sun, Y. Xie, Y. Wang, Y. Yang, C. Tian, L. Wang, H. Fu, Angew. Chem. Int. Ed. **60**, 14005 (2021)
236. Z. Wang, C. Zhu, H. Tan, J. Liu, L. Xu, Y. Zhang, Y. Liu, X. Zou, Z. Liu, X. Lu, Adv. Funct. Mater. **31**, 2104735 (2021)
237. S.-H. Yin, J. Yang, Y. Han, G. Li, L.-Y. Wan, Y.-H. Chen, C. Chen, X.-M. Qu, Y.-X. Jiang, S.-G. Sun, Angew. Chem. Int. Ed. **59**, 21976 (2020)
238. Y. Hou, M. Qiu, M.G. Kim, P. Liu, G. Nam, T. Zhang, X. Zhuang, B. Yang, J. Cho, M. Chen, C. Yuan, L. Lei, X. Feng, Nat. Commun. **10**, 1392 (2019)
239. Z. Zhang, X. Zhao, S. Xi, L. Zhang, Z. Chen, Z. Zeng, M. Huang, H. Yang, B. Liu, S.J. Pennycook, P. Chen, Adv. Energy Mater. **10**, 2002896 (2020)
240. Y. Zhou, X. Tao, G. Chen, R. Lu, D. Wang, M.-X. Chen, E. Jin, J. Yang, H.-W. Liang, Y. Zhao, X. Feng, A. Narita, K. Müllen, Nat. Commun. **11**, 5892 (2020)
241. M. Xiao, Z. Xing, Z. Jin, C. Liu, J. Ge, J. Zhu, Y. Wang, X. Zhao, Z. Chen, Adv. Mater. **32**, 2004900 (2020)
242. J. Ban, X. Wen, H. Xu, Z. Wang, X. Liu, G. Cao, G. Shao, J. Hu, Adv. Funct. Mater. **31**, 2010472 (2021)
243. R. Lang, X. Du, Y. Huang, X. Jiang, Q. Zhang, Y. Guo, K. Liu, B. Qiao, A. Wang, T. Zhang, Chem. Rev. **120**, 11986 (2020)
244. G. Giannakakis, M. Flytzani-Stephanopoulos, E.C.H. Sykes, Acc. Chem. Res. **52**, 237 (2019)
245. Y. Xu, Y. Zhang, Z. Guo, J. Ren, Y. Wang, H. Peng, Angew. Chem. Int. Ed. **54**, 15390 (2015)
246. Z. Guo, C. Li, W. Li, H. Guo, X. Su, P. He, Y. Wang, Y. Xia, J. Mater. Chem. A **4**, 6282 (2016)
247. Z.-Y. Chen, Y.-N. Li, L.-L. Lei, S.-J. Bao, M.-Q. Wang, L. Heng, Z.-L. Zhao, M.-W. Xu, Catal. Sci. Technol. **7**, 5670 (2017)
248. H.-F. Wang, C. Tang, B. Wang, B.-Q. Li, X. Cui, Q. Zhang, Energy Storage Mater. **15**, 124 (2018)
249. Q. Liu, Y. Wang, L. Dai, J. Yao, Adv. Mater. **28**, 3000 (2016)
250. Y. Chen, H. Wang, S. Ji, B.G. Pollet, R. Wang, J. Ind. Eng. Chem. **71**, 284 (2019)
251. J. Zhang, Z. Zhao, Z. Xia, L. Dai, Nat. Nanotechnol. **10**, 444 (2015)
252. H. Yu, H. Zhang, Z. Zhang, ChemCatChem **13**, 397 (2021)
253. X. Liu, L. Dai, Nat. Rev. Mater. **1**, 16064 (2016)
254. Z. W. Seh, J. Kibsgaard, C. F. Dickens, I. Chorkendorff, J. K. Nørskov, T. F. Jaramillo, Science, **355**, eaad4998 (2017)
255. M. Wu, J. Shi, Q. Wang, J. Qiao, Y. Liu, ECS Trans. **66**, 79 (2015)
256. H. Jiang, J. Gu, X. Zheng, M. Liu, X. Qiu, L. Wang, W. Li, Z. Chen, X. Ji, J. Li, Energy Environ. Sci. **12**, 322 (2019)
257. C. Hang, J. Zhang, J. Zhu, W. Li, Z. Kou, Y. Huang, Adv. Energy Mater. **8**, 1703539 (2018)
258. J. Han, G. Huang, Z. Wang, Z. Lu, J. Du, H. Kashani, M. Chen, Adv. Mater. **30**, 1803588 (2018)

259. H.B. Yang, J. Miao, S.-F. Hung, J. Chen, H.B. Tao, X. Wang, L. Zhang, R. Chen, J. Gao, H.M. Chen, L. Dai, B. Liu, Sci. Adv. **2**, e1501122 (2016)
260. X. Xiao, X. Li, Z. Wang, G. Yan, H. Guo, Q. Hu, L. Li, Y. Liu, J. Wang, Appl. Catal., B, **265**, 118603 (2020)
261. Z. Zhao, M. Li, L. Zhang, L. Dai, Z. Xia, Adv. Mater. **27**, 6834 (2015)
262. K. Sheng, Q. Yi, L. Hou, A. Chen, J. Electrochem. Soc. **167**, 070560 (2020)
263. L.-L. Ma, W.-J. Liu, X. Hu, P.K.S. Lam, J.R. Zeng, H.-Q. Yu, Chem. Eng. J. **400**, 125969 (2020)
264. X. Hao, W. Chen, Z. Jiang, X. Tian, X. Hao, T. Maiyalagan, Z.-J. Jiang, Electrochim. Acta **362**, 137143 (2020)
265. H. Zheng, Y. Zhang, J. Long, R. Li, X. Gou, J. Electrochem. Soc. **167**, 084516 (2020)
266. J.-J. Cai, Q.-Y. Zhou, X.-F. Gong, B. Liu, Y.-L. Zhang, Y.-K. Dai, D.-M. Gu, L. Zhao, X.-L. Sui, Z.-B. Wang, Carbon **167**, 75 (2020)
267. H. Ji, M. Wang, S. Liu, H. Sun, J. Liu, T. Qian, C. Yan, Electrochim. Acta **334**, 135562 (2020)
268. Q. Lv, N. Wang, W. Si, Z. Hou, X. Li, X. Wang, F. Zhao, Z. Yang, Y. Zhang, C. Huang, Appl. Catal., B, **261**, 118234 (2020)
269. K. Gong, F. Du, Z. Xia, M. Durstock, L. Dai, Science **323**, 760 (2009)
270. J.T. Ren, G.G. Yuan, C.C. Weng, L. Chen, Z.Y. Yuan, ChemCatChem **10**, 5297 (2018)
271. S. Yi, X. Qin, C. Liang, J. Li, R. Rajagopalan, Z. Zhang, J. Song, Y. Tang, F. Cheng, H. Wang, M. Shao, Appl. Catal., B, **264**, 118537 (2020)
272. M. Wu, Y. Wang, Z. Wei, L. Wang, M. Zhuo, J. Zhang, X. Han, J. Ma, J. Mater. Chem. A **6**, 10918 (2018)
273. J. Tong, W. Ma, L. Bo, T. Li, W. Li, Y. Li, Q. Zhang, J. Power Sources **441**, 227166 (2019)
274. Z. Zhao, Z. Yuan, Z. Fang, J. Jian, J. Li, M. Yang, C. Mo, Y. Zhang, X. Hu, P. Li, S. Wang, W. Hong, Z. Zheng, G. Ouyang, X. Chen, D. Yu, Adv. Sci. **5**, 1800760 (2018)
275. M. Wu, J. Qiao, K. Li, X. Zhou, Y. Liu, J. Zhang, Green Chem. **18**, 2699 (2016)
276. C. Tang, M.M. Titirici, Q. Zhang, J. Energy Chem. **26**, 1077 (2017)
277. X. Xue, H. Yang, T. Yang, P. Yuan, Q. Li, S. Mu, X. Zheng, L. Chi, J. Zhu, Y. Li, J. Zhang, Q. Xu, J. Mater. Chem. A **7**, 15271 (2019)
278. S. Chen, L. Zhao, J. Ma, Y. Wang, L. Dai, J. Zhang, Nano Energy **60**, 536 (2019)
279. J. Guo, Y. Yu, J. Ma, T. Zhang, S. Xing, J. Alloys Compd. **821**, 153484 (2020)
280. M. Wu, Q. Tang, F. Dong, Z. Bai, L. Zhang, J. Qiao, J. Catal. **352**, 208 (2017)
281. Z. Pei, Z. Li, Y. Huang, Q. Xue, Y. Huang, M. Zhu, Z. Wang, C. Zhi, Energy Environ. Sci. **10**, 742 (2017)
282. M. Wu, G. Zhang, M. Wu, J. Prakash, S. Sun, Energy Storage Mater. **21**, 253 (2019)
283. C.H. Choi, H.K. Lim, M.W. Chung, J.C. Park, H. Shin, H. Kim, S.I. Woo, J. Am. Chem. Soc. **136**, 9070 (2014)
284. C. Hu, D. Liu, Y. Xiao, L. Dai, Prog. Nat. Sci.: Mater. Int. **28**, 121 (2018)
285. F. Dong, Y. Cai, C. Liu, J. Liu, J. Qiao, Int. J. Hydrogen Energy **43**, 12661 (2018)
286. L. Wang, Y. Wang, M. Wu, Z. Wei, C. Cui, M. Mao, J. Zhang, X. Han, Q. Liu, J. Ma, Small **14**, 1800737 (2018)
287. M. Wu, G. Zhang, Y. Hu, J. Wang, T. Sun, T. Regier, J. Qiao, S. Sun, Carbon Energy **3**, 176 (2021)
288. D. Ji, L. Fan, L. Tao, Y. Sun, M. Li, G. Yang, T.Q. Tran, S. Ramakrishna, S. Guo, Angew. Chem. Int. Ed. **58**, 13840 (2019)
289. N. Xu, Q. Nie, J. Liu, H. Huang, J. Qiao, X.-D. Zhou, J. Electrochem. Soc. **167**, 050512 (2020)
290. P. Yan, M. Huang, B. Wang, Z. Wan, M. Qian, H. Yan, T.T. Isimjan, J. Tian, X. Yang, J. Energy Chem. **47**, 299 (2020)
291. Z. Ma, S. Dou, A. Shen, L. Tao, L. Dai, S. Wang, Angew. Chem. Int. Ed. **2015**, 54 (1888)
292. Y. Wang, T. Zhou, K. Jiang, P. Da, Z. Peng, J. Tang, B. Kong, W.-B. Cai, Z. Yang, G. Zheng, Adv. Energy Mater. **4**, 1400696 (2014)
293. K. Tao, J.O. Donnell, H. Yuan, E.U. Haq, S. Guerin, L.J.W. Shimon, B. Xue, C. Silien, Y. Cao, D. Thompson, R. Yang, S.A.M. Tofail, E. Gazit, Energy Environ. Sci. **13**, 96 (2020)

294. Z. Li, L. Lv, X. Ao, J.-G. Li, H. Sun, P. An, X. Xue, Y. Li, M. Liu, C. Wang, M. Liu, Appl. Catal., B, **262**, 118291 (2020)
295. G.S. Hegde, A. Ghosh, R. Badam, N. Matsumi, R. Sundara, A.C.S. Appl, Energy Mater. **3**, 1338 (2020)
296. C. Li, M. Wu, R. Liu, Appl. Catal., B, **244**, 150 (2019)
297. Y. Wang, T. Hu, Y. Qiao, Y. Chen, Int. J. Hydrogen Energy **45**, 8686 (2020)
298. W. Niu, Z. Li, K. Marcus, L. Zhou, Y. Li, R. Ye, K. Liang, Y. Yang, Adv. Energy Mater. **8**, 1701642 (2018)
299. D. Zhao, Z. Tang, W. Xu, Z. Wu, L.J. Ma, Z. Cui, C. Yang, L. Li, J. Colloid Interface Sci. **560**, 186 (2020)
300. B. Liu, S. Qu, Y. Kou, Z. Liu, X. Chen, Y. Wu, X. Han, Y. Deng, W. Hu, C. Zhong, A.C.S. Appl, Mater. Interfaces **10**, 30433 (2018)
301. J. Zhu, T. Qu, F. Su, Y. Wu, Y. Kang, K. Chen, Y. Yao, W. Ma, B. Yang, Y. Dai, F. Liang, D. Xue, Dalton Trans. **2020**, 49 (1811)
302. F. Meng, H. Zhong, D. Bao, J. Yan, X. Zhang, J. Am. Chem. Soc. **138**, 10226 (2016)
303. M. Wu, G. Zhang, J. Qiao, N. Chen, W. Chen, S. Sun, Nano Energy **61**, 86 (2019)
304. S. Gupta, L. Qiao, S. Zhao, H. Xu, Y. Lin, S.V. Devaguptapu, X. Wang, M.T. Swihart, G. Wu, Adv. Energy Mater. **6**, 1601198 (2016)
305. M. Wu, Q. Wei, G. Zhang, J. Qiao, M. Wu, J. Zhang, Q. Gong, S. Sun, Adv. Energy Mater. **8**, 1801836 (2018)
306. N. Huang, S. Yan, L. Yang, M. Zhang, P. Sun, X. Lv, X. Sun, J. Solid State Chem. **285**, 121185 (2020)
307. Y. Guo, P. Yuan, J. Zhang, H. Xia, F. Cheng, M. Zhou, J. Li, Y. Qiao, S. Mu, Q. Xu, Adv. Funct. Mater. **28**, 1805641 (2018)
308. J. Song, C. Zhu, S. Fu, Y. Song, D. Du, Y. Lin, J. Mater. Chem. A **4**, 4864 (2016)
309. M. Zeng, Y. Liu, F. Zhao, K. Nie, N. Han, X. Wang, W. Huang, X. Song, J. Zhong, Y. Li, Adv. Funct. Mater. **26**, 4397 (2016)
310. G. Nam, J. Park, M. Choi, P. Oh, S. Park, M.G. Kim, N. Park, J. Cho, J.-S. Lee, ACS Nano **9**, 6493 (2015)
311. S. Gadipelli, T. Zhao, S.A. Shevlin, Z. Guo, Energy Environ. Sci. **9**, 1661 (2016)
312. Y. Liu, F. Chen, W. Ye, M. Zeng, N. Han, F. Zhao, X. Wang, Y. Li, Adv. Funct. Mater. **27**, 1606034 (2017)
313. S. Liu, Z. Wang, S. Zhou, F. Yu, M. Yu, C.Y. Chiang, W. Zhou, J. Zhao, J. Qiu, Adv. Mater. **29**, 1700874 (2017)
314. G. Fu, Y. Chen, Z. Cui, Y. Li, W. Zhou, S. Xin, Y. Tang, J.B. Goodenough, Nano Lett. **16**, 6516 (2016)
315. J. Zhu, M. Xiao, Y. Zhang, Z. Jin, Z. Peng, C. Liu, S. Chen, J. Ge, W. Xing, ACS Catal. **6**, 6335 (2016)
316. H. Furukawa, K.E. Cordova, M. O'Keeffe, O.M. Yaghi, Science **341**, 1230444 (2013)
317. J. Tang, R.R. Salunkhe, J. Liu, N.L. Torad, M. Imura, S. Furukawa, Y. Yamauchi, J. Am. Chem. Soc. **137**, 1572 (2015)
318. W. Chaikittisilp, N. L. Torad, C. Li, M. Imura, N. Suzuki, S. Ishihara, K. Ariga, Y. Yamauchi, Chem. Eur. J., **20**, 4217 (2014)
319. M. Wang, T. Qian, J. Zhou, C. Yan, A.C.S. Appl, Mater. Interfaces **9**, 5213 (2017)
320. B.Y. Guan, L. Yu, X.W. Lou, Energy Environ. Sci. **9**, 3092 (2016)
321. S. Zhao, H. Yin, L. Du, L. He, K. Zhao, L. Chang, G. Yin, H. Zhao, S. Liu, Z. Tang, ACS Nano **8**, 12660 (2014)
322. J.-S. Lee, G. Nam, J. Sun, S. Higashi, H.-W. Lee, S. Lee, W. Chen, Y. Cui, J. Cho, Adv. Energy Mater. **6**, 1601052 (2016)
323. W. Wang, L. Kuai, W. Cao, M. Huttula, S. Ollikkala, T. Ahopelto, A. Honkanen, S. Huotari, M. Yu, B. Geng, Angew. Chem., Int. Ed. **56**, 14977 (2017)
324. D. Lee, M. Park, Z. Cano, W. Ahn, Z. Chen, ChemSusChem **11**, 406 (2018)
325. P. Menezes, A. Indra, N. Sahraie, A. Bergmann, P. Strasser, M. Driess, ChemSusChem **8**, 164 (2015)

326. A. Aijaz, J. Masa, C. Rosler, W. Xia, P. Weide, A. Botz, R. Fischer, W. Schuhmann, M. Muhler, Angew. Chem., Int. Ed. **55**, 4087 (2016)
327. S. Devaguptapu, S. Hwang, S. Karakalos, S. Zhao, S. Gupta, D. Su, H. Xu, G. Wu, ACS Appl. Mater. Interfaces **9**, 44567 (2017)
328. H. Cheng, M. Li, C. Su, N. Li, Z. Liu, Adv. Funct. Mater. **27**, 1701833 (2017)
329. Z. Liu, H. Cheng, N. Li, T. Ma, Y. Su, Adv. Mater. **28**, 3777 (2016)
330. C. Jin, F. Lu, X. Cao, Z. Yang, R. Yang, J. Mater. Chem. A **1**, 12170 (2013)
331. K. Chakrapani, G. Bendt, H. Hajiyani, T. Lunkenbein, M. Greiner, L. Masliuk, S. Salamon, J. Landers, R. Schlögl, H. Wende, R. Pentcheva, S. Schulz, M. Behrens, ACS Catal. **8**, 1259 (2018)
332. W. Song, Z. Ren, S. Chen, Y. Meng, S. Biswas, P. Nandi, H. Elsen, P. Gao, S. Suib, ACS Appl. Mater. Interfaces **8**, 20802 (2016)
333. J. Bian, X. Cheng, X. Meng, J. Wang, J. Zhou, S. Li, Y. Zhang, C. Sun, ACS Appl. Energy Mater. **2**, 2296 (2019)
334. Z. Li, Y. Zhang, Y. Feng, C. Cheng, K. Qiu, C. Dong, H. Liu, X. Du, Adv. Funct. Mater. **29**, 1903444 (2019)
335. W. Liu, J. Bao, L. Xu, M. Guan, Z. Wang, J. Qiu, Y. Huang, J. Xia, Y. Lei, H. Li, Appl. Surf. Sci. **478**, 552 (2019)
336. Q. Wen, C. Xi, Y. Zhang, R. Zhang, Z. Li, G. Sheng, H. Liu, C. Dong, Y. Chen, X. Du, Chem. Commun. **55**, 8579 (2019)
337. Y. Su, H. Liu, C. Li, J. Liu, Y. Song, F. Wang, J. Alloys Compd. **799**, 160 (2019)
338. X. Cao, W. Yan, C. Jin, J. Tian, K. Ke, R. Yang, Electrochim. Acta **180**, 788 (2015)
339. X. Han, G. He, Y. He, J. Zhang, X. Zheng, L. Li, C. Zhong, W. Hu, Y. Deng, T. Ma, Adv. Energy Mater. **8**, 1702222 (2018)
340. X. He, F. Yin, Y. Li, H. Wang, J. Chen, Y. Wang, B. Chen, ACS Appl. Mater. Interfaces **8**, 26740 (2016)
341. S. Cui, L. Sun, F. Kong, L. Huo, H. Zhao, J. Power Sources **430**, 25 (2019)
342. T. Lippert, M. Montenegro, M. Döbeli, A. Weidenkaff, S. Müller, P. Willmott, A. Wokaun, Prog. Solid State Chem. **35**, 221 (2007)
343. Y. Zhao, L. Xu, L. Mai, C. Han, Q. An, X. Xu, X. Liu, Q. Zhang, Proc. Natl. Acad. Sci. USA **109**, 19569 (2012)
344. J. Jung, M. Risch, S. Park, M. Kim, G. Nam, H. Jeong, S. Yang, J. Cho, Energy Environ. Sci. **9**, 176 (2016)
345. G. Liu, H. Chen, L. Xia, S. Wang, L. Ding, D. Li, K. Xiao, S. Dai, H. Wang, ACS Appl. Mater. Interfaces **7**, 22478 (2015)
346. J. Zhang, C. Zhang, W. Li, Q. Guo, H. Gao, Y. You, Y. Li, Z. Cui, K. Jiang, H. Long, D. Zhang, S. Xin, ACS Appl. Mater. Interfaces **10**, 5543 (2018)
347. Z. Wang, C. Jin, J. Sui, C. Li, R. Yang, Int. J. Hydrogen Energy **43**, 20727 (2018)
348. S. Peng, X. Han, L. Li, S. Chou, D. Ji, H. Huang, Y. Du, J. Liu, S. Ramakrishna, Adv. Energy Mater. **8**, 1800612 (2018)
349. Z. Wu, L. Sun, T. Xia, L. Huo, H. Zhao, A. Rougier, J. Grenier, J. Power Sources **334**, 86 (2016)
350. P. Li, B. Wei, Z. Lü, Y. Wu, Y. Zhang, X. Huang, Appl. Surf. Sci. **464**, 494 (2019)
351. B. Hua, Y. Sun, M. Li, N. Yan, J. Chen, Y. Zhang, Y. Zeng, B. Amirkhiz, J. Luo, Chem. Mater. **29**, 6228 (2017)
352. J. Wang, H. Zhao, Y. Gao, D. Chen, C. Chen, M. Saccoccio, F. Ciucci, Int. J. Hydrogen Energy **41**, 10744 (2016)
353. P. Da, M. Wu, K. Qiu, D. Yan, Y. Li, J. Mao, C. Dong, T. Ling, S. Qiao, Chem. Eng. Sci. 194, 127 (2019)

Chapter 4
Anode of Zn-Air Battery

4.1 Background on Zn Electrodes

Zn-air batteries are renewable energy devices that generate electrical energy through a redox reaction between the anode metal and the cathode oxygen. In recent years, rechargeable Zn-air batteries have been extensively studied for their great potential in the consumer electronics market and the field of mobile power. They offer greater advantages in many fields than other batteries. For example, metal ion batteries, particularly lithium-ion batteries, are widely used in electronic devices due to the high open-circuit voltage, large power density, long operating life, and relatively advantageous energy density [1, 2]. In addition, proton exchange membrane fuel cells (PEMFC) offer advantages in terms of energy and power density [3]. Considering their operating life, manufacturing costs, energy density, operational safety, etc., rechargeable Zn-air batteries offer significant advantages as a possible alternative in this respect. For example, high energy density (1353 Wh kg^{-1}), low cost (currently < \$100 kWh^{-1} and likely to be reduced to < \$10 kWh^{-1} in the future), environmental friendliness and excellent safety make them another promising candidate for next-generation advanced energy devices [4–6].

Its structural configuration is shown in Fig. 4.1, and the cathode electrode (including electrocatalyst, air electrode, and collector), electrolyte, separator, and anode zinc electrode constitute the main components of the Zn-air rechargeable battery. During the discharging process, the zinc electrode oxidizes, while oxygen at the three-phase interface of the air electrode participates in the reduction reaction and generates OH$^-$. The zinc oxide on the electrode is reduced to metallic zinc, and OH$^-$ is oxidized to form oxygen at the three-phase interface of the air electrode during the charging process.

The charging and discharging reactions of rechargeable Zn-air batteries are reversible, and their reactions can be summarized as follows:

The discharging process:

Air electrode reaction:

© The Author(s), under exclusive license to Springer Nature Singapore Pte Ltd. 2023
S. Peng, *Zinc-Air Batteries*, https://doi.org/10.1007/978-981-19-8214-9_4

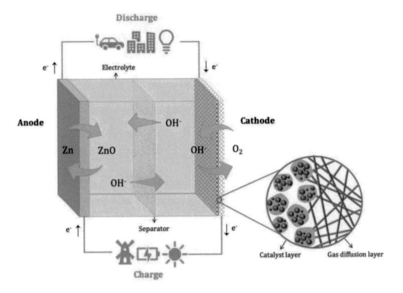

Fig. 4.1 Diagram of a zinc-air battery installation

$$O_2 + 4e^- + 2H_2O \rightarrow 4OH^- \, E = 0.40 \text{ V versus SHE}$$

Zinc electrode reaction:

$$Zn + 4OH^- \rightarrow Zn(OH)_4^{2-} + 2e^-$$
$$Zn(OH)_4^{2-} \rightarrow ZnO + H_2O + 2OH^- \, E = 1.26 \text{ V versus SHE}$$

The charging process:

Air electrode reaction:

$$4OH^- \rightarrow O_2 + 4e^- + 2H_2O$$

Zinc electrode reaction:

$$ZnO + H_2O + 2e^- \rightarrow Zn + 2OH$$

Overall reaction:

$$2Zn + O_2 \rightarrow 2ZnO \, E = 1.66 \text{ V versus SHE}$$

This shows that the anode and the surface/interface are the main reaction sites in rechargeable Zn-air batteries and their performance is closely related to the performance of the whole battery [7–9]. Over the past decade, researchers have explored several low-cost, high-capacity anode and cathode materials for the development of electrically rechargeable Zinc-air batteries [10, 11]. However, the air and zinc electrodes of secondary Zn-air batteries still face several key challenges that impede the ac process. Many excellent comments have focused on the challenges faced in the development of air electrodes, such as finding alternatives to carbon or developing new strategies to achieve efficient dual-function catalysts that effectively reduce discharge/charge overpotential [12–14]. In contrast, the rather limited review work has focused comprehensively on the challenges facing zinc anodes. Zinc electrodes are a more serious constraint on the performance of Zn-air batteries: cell failure is usually attributed to the degradation of the zinc anode rather than the air electrode, since the air electrode usually has a much longer cycle life than the zinc anode. Many studies show that when zinc electrode degrades, air electrode does not degenerate at all [15–17]. Therefore, the problem of zinc anode is more urgent to be solved at present in the battery.

The key problems of zinc electrodes are passivation, dendrite growth, and hydrogen precipitation reaction (Fig. 4.2). By weakening the discharge performance, the shelf life and cycle life of rechargeable Zinc-air batteries are reduced, which limits the practical application [18–20]. The main purpose of this view is to summarize the zinc anode issues that limit the commercial application of rechargeable Zn-air batteries and to guide future academic research based on current research on existing challenges.

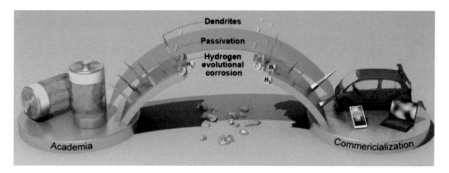

Fig. 4.2 Problems with the cathode of zinc-air batteries

4.1.1 The Anode Passivation

In practice, the actual energy density of the Zn-air battery is 220–300 Wh kg^{-1}, far behind the theoretical value (1086 Wh kg^{-1}), which remains an unattainable challenge and requires significant efforts for its development [21]. Due to the passivation of zinc electrodes in alkaline electrolytes, the utilization of zinc is low, which is one of the important reasons for the discrepancy between theoretical and practical energy densities. A typical redox reaction of zinc anode in alkaline water-electrolyte can be described as follows [22]:

$$Zn(s) + 4OH^- \rightarrow Zn(OH)_4^{2-}(aq) + 2e^- \tag{4.1}$$

$$Zn(OH)_4^{2-} \rightarrow ZnO(s) + 2OH^- + H_2O \tag{4.2}$$

Passivation refers to the supersaturation of zinc oxide in the alkaline electrolyte, and it is deposited on the surface of the zinc electrode to form an insulating layer during the discharge process, which blocks the surface of the electrode and affects its further reaction. Insulating zinc oxide passivation film terminates the discharge process, reducing zinc electrode utilization and battery capacity. It also prevents the reverse conversion of metallic zinc, limiting the battery rechargeability (Fig. 4.3a, b) [23].

Few research has focused on the mechanism of passivation. Liu et al. believed that the total passivation time (t) of zinc in alkaline solution was the sum of the saturation time of zincate (t_a), the formation time of porous zinc oxide layer (t_b), and the formation time of dense Zinc oxide layer (t_c) [26]. At high current densities, dense zinc oxide layers form passivation electrodes. At low current densities, a porous zinc oxide layer is formed, and passivation does not occur when the mass transfer of hydroxide ions on the porous zinc oxide layer is sufficient for anodization. Sinu et al. suggested that the passivation layer was divided into two parts: the inner layer was a dense zinc oxide layer, while the outer layer was mainly composed of ZnO/Zn(OH)$_2$ sediments. The mechanism of passivation still needs a deep and unified understanding [27].

The performance of the Zn-air battery depends not only on the zinc electrode, but also strongly on the interaction between the electrode and electrolyte. The composition and concentration of electrolytes affect the formation process of zinc oxide, and further affect the discharge capacity and rechargeable battery. Therefore, to reduce the influence of passivation on the battery charge–discharge performance, zinc passivation of Zn-air batteries can be eliminated or reduced by designing metal anode materials and changing the composition of electrolyte.

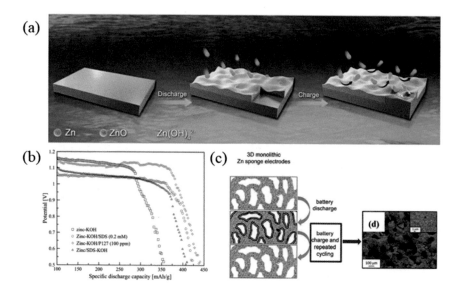

Fig. 4.3 a Schematic diagram of passivation layer formation during Zn metal anode discharge. The passivation layer is not fully reversible during charging: and appears as a thin layer of zinc oxide after charging. Indicates that there is still a thin layer of zinc oxide after charging. **b** Electrostatic discharge curves of batteries with an electrolyte circulation rate of 150 mL min^{-1} and discharge current density of 25 mA cm^{-2} [23]. **c** Schematic cross-section of the Zn sponge electrode for 3D wiring [24]. **d** Microscopic analysis of Zn sponge after about 54,000 cycles [25]

4.1.2 Dendrite Growth

Like other metal–oxygen cell systems with metal anodes (such as Na or Li), alkaline Zn-air cells generally form dendrites on the anode surface during charging. The mechanism of zinc dendrite formation has been studied. The charging reaction of the zinc electrode is mainly affected by liquid mass transfer, and a large concentration polarization is formed in the low concentration of active components near the surface of zinc electrode. The active components in the electrolyte are more likely to diffuse on the electrode surface, where the current density of the reaction is much higher, resulting in an uneven distribution of current across the electrode [28, 29]. Eventually, the deposition of zinc at the tip of the electrode accelerates, forming the final dendrites [30]. During the charging process, the deposition morphology of zinc electrodes changes with the change of current density. Due to low activation overpotential, dendritic morphology does not form at low current density. It is easy to form leaf-like morphology under high current density, indicating that the twigs grow symmetrically on both sides of the main stem. This fractal growth may result from strong surface energy anisotropy, which limits the diffusion of atoms. In general, dendritic crystals are axial and needle-like single crystals with flat stepped tops, as shown in Fig. 4.5a [31]. Growing twigs can lead to short circuits and deteriorating electrochemical performance, resulting in shorter battery life [32]. In addition, the twigs can easily fall off the electrode, leading to a decline in battery capacity.

4.1.3 Hydrogen Evolution Reaction

The batteries mentioned above are sometimes idle or low-power in practical applications. Therefore, in addition to cycle life, battery shelf life is also an important performance evaluation index. Metallic zinc is thermodynamically unstable in alkaline solutions because zinc has a more negative reduction potential than hydrogen, leading to the evolution of hydrogen. Hydrogen evolution reaction results in serious performance degradation of Zn-air batteries, which poses a threat to shelf life. The hydrogen evolutionary reaction (HER) is shown below [33, 34].

$$Zn + H_2O \rightarrow H_2 + Zn(OH)_2 \tag{4.3}$$

Hydrogen can be formed by chemical and electrochemical reactions between zinc and water during electrodeposition. In particular, hydrogen is formed mainly through the corrosion of zinc electrodes, rather than through the electrochemical formation of zinc in the electrochemical reduction process. On the one hand, the hydrogen produced increases the internal pressure of the battery, damaging it as it expands. On the other hand, HER reaction consumes the active metal zinc, leading to the decay of battery capacity and self-discharge. Over long periods, the products of hydrogen evolution reactions may also lead to blocking effects similar to passivation.

4.2 The Research Status

4.2.1 Modified Electrolyte Components for Passivation

The zinc electrode is in close contact with electrolyte; therefore, the passivation of zinc electrode can be inhibited by modifying the composition of electrolyte. In general, several methods have been developed to inhibit the passivation layer of zinc oxide, such as increasing the concentration of the electrolyte, adding surfactants to the electrolyte, and creating a flow of liquid to dissolve or destroy the dense structure of the passivation layer. Among these methods, it is the easiest to use a high concentration of alkaline electrolyte to increase the solubility of the accumulated zinc oxide, since the passivation layer is loose and unstable at high pH, but this method can also cause severe anodic corrosion. Adding an appropriate amount of surfactant in the electrolyte can also prevent zinc oxide precipitation on the surface of active zinc and make the passivation layer lose, to reduce the influence of the passivation layer on the ability of zinc ions generated by the discharge reaction to diffuse into the electrolyte, and reduce the diffusion resistance of passivation layer on the discharge products [35]. Surfactants are amphiphilic molecules that contain both hydrophobic and hydrophilic groups. Various types of surfactants have been used in different battery systems to improve their electrochemical performance.

Surfactants commonly used in Zn-air batteries can be divided into three categories. (i) Cationic surfactants such as dodecyltrimethylammonium bromide (DTAB) and cetyltrimethylammonium bromide (CTAB) [36], (ii) Anionic surfactants such as sodium dodecyl sulfate (SDS) and sodium dodecyl benzene sulfonate (SDBS) [35, 37, 38], and (iii) non-ionic surfactants such as polyoxyethylene nonylbenzene and Pluronic F-127 (P127) [39]. These surfactants significantly inhibit the passivation and corrosion of zinc anodes in dilute alkaline electrolytes because the adsorbed surfactant molecules prevent contact between water and anode. Some studies have shown that anionic surfactants (such as SDBS) are superior to cationic surfactants (such as CTAB) in improving battery cycle life [40, 41], The negatively charged polar groups of SDBS coordinate with zinc ions on the anode surface, resulting in the formation of small zinc oxide particles, which effectively inhibit the growth of the passivation layer. In contrast, the positively charged polar groups of CTAB show weak interactions with zinc ions, resulting in the formation of larger zinc oxide particles. In addition, surfactants with multiple anchoring groups have an advantage in binding to the anode surface to inhibit the formation of zinc oxide passivation layers. Figure 4.3b shows that the discharge capacity of the cell using the P127 additive is significantly better than that using the SDS additive. The reason is that each molecule of P127 has multiple anchoring groups, which have good interaction and strong adsorption capacity with the substrate. In contrast, SDS has an anchor group and shows weak surface adsorption [42].

In addition, for Zn-air flow cells, the convective condition of electrolytes also affects the physical and chemical properties of zinc passivation film [43]. The flowing electrolyte flow reduces the effect of the passivation layer by removing the loosely deposited passivation layer and reducing the local concentration gradient. However, most studies on the modification of electrolyte composition focus only on inhibiting the formation of zinc oxide passivation layer during discharge. Rechargeability in electric charging process has not been paid attention to. Although the passivation layer may reduce the hindrance to reverse conversion of metal zinc, the problem of anodic dissolution still exists.

4.2.2 Preparation of Modified Anodes for Passivation Suppression

The development of a modified metal anode is an effective means to improve battery capacity and charging capacity. The preparation of modified anode includes adjustment of zinc anode structure and surface modification of anode particles. The adjustment of zinc anode structure refers to the development of porous active anode material with a high surface area. The surface modification of nanocrystalline anode particles refers to coating zinc anode particles with various ion exchange nanoshells.

The development of active anode material with sufficient porosity and effective specific surface area by pulse electroplating can improve the corrosion sensitivity of anode material and inhibit the formation of a passivation layer [44]. Therefore, the utilization rate, discharge capacity, and energy density of anode materials are improved. It should be noted that when the specific surface area of the anode material is overextended, negative effects such as increased electrode resistance and corrosion rate can shorten its storage life [45, 46]. The zinc electrode can also be heat-treated to form a spongy porous structure, which allows for high specific capacity and enhanced charging capacity. As shown in Fig. 4.3c, the integral, porous, and aperiodic structure allows deep discharge of the inner core of the electrically conductive metal zinc. The cell can be cycled hundreds to thousands of times without passivation or dendrite (Fig. 4.4c). However, the characteristic size of the microporous zinc sponge is $\sim 10\ \mu$ ($2\ \mu$ above the critical passivation size), so there is room to improve the discharge depth (DOD) [47].

Sub-micron anode particles with nanometer diameter (< 500 nm) are also being studied because they can simultaneously prevent passivation and keep the material active during long cycles. However, due to the large pole-electrolyte surface area of these submicron zinc anodes, the resulting anodic dissolution problem is not well controlled or clearly explained [48]. Submicron-sized anodes will exhibit severe morphological degradation as almost all anode particles fall off the substrate due to dissolution, resulting in poor cycling performance [49]. The problem of passivation and dissolution can be alleviated by coating nano-ZnO particles with ion-exchange nanoshells, which has aroused great commercial and research interest in rechargeable Zn-air batteries. The aperture of the coated nanoshell is customized to allow the passage of hydroxide ions while blocking the transfer of zincate ions (Fig. 4.4a). During charging, the zincate intermediate gets trapped inside the carbon shell and reacts with the zinc inside the shell, preventing zinc from depositing in another location. By contrast, due to its small size, OH^- by-products can be freely diffused out through the micropores and H_2O in the shell. During the discharge process, the trapped zinc is oxidized to form ZnO with the participation of OH^-/H_2O from outside the shell. To create an ideal environment for free electron transport and thus the electrochemical activity of zinc oxide, ion-isolated nanocapsid components such as GO, TiN_xO_y coating [49], carbon coating [23], silicon-based coating [50], and ionomer hydroxides conductive polymer [51] need to be selected. Figure 4.4b shows that the life of the carbon-coated zinc oxide (ZnO@C) anode exceeds that of the bare zinc oxide anode by about 1.6 times in terms of cycles. In particular, the cycle starting with zinc oxide is more feasible because it avoids the formation of cracks and cracks caused by the expansion of the active material during the cycling. SEM images of the exposed and coated zinc oxide anodes (Fig. 4.4c) show that the surface of the exposed zinc oxide anodes has holes after circulation, while the ZnO@C anodes retain their electrode morphology, confirming that the carbon shell on ZnO@C can alleviate the dissolution and passivation of zinc anodes [23]. However, the study must be concerned with performance maintenance over longer operating times and deeper

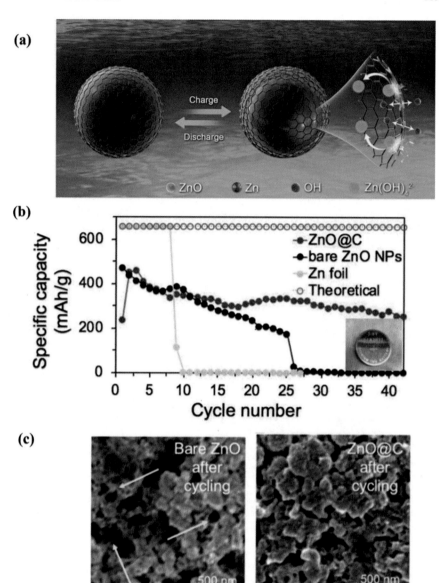

Fig. 4.4 a Schematic design of ion exchange nanoshell for zinc anode. **b** Specific capacity and Coulomb efficiency of bare ZnO (1.03 mg) and ZnO@C (0.94 mg). **c** SEM images of a bare zinc anode and ZnO@C anode after three cycles [47]

(a)

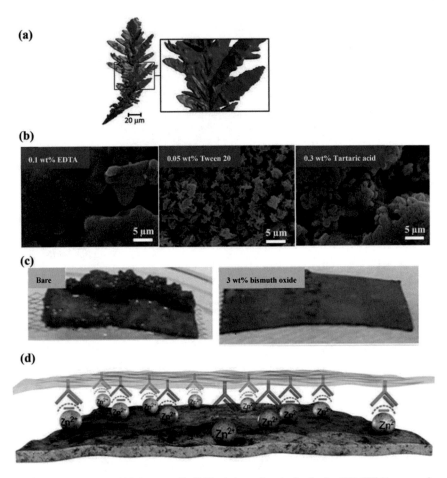

(b)

(c)

(d)

Fig. 4.5 **a** Reconstructed 3D images of individual zinc twigs obtained using FIB-SEM tomography [31]. **b** The SEM images of Zinc electrodes with different organic additives after the 1000th cycle [55]. **c** Zinc anode after the 20th cycle of inorganic additives [56]. **d** Sketch of quasi-SEI formation between zinc anode and PANa electrolyte [57]

DOD, as the carbon-coated anodes may have weak mechanical properties when the middle empty region (the region filled by what is in Fig. 4.4a) is inevitably formed by the volume shrinkage of zinc to ZnO. The zinc anode (Zn-POME) with a garnet structure is coated by a conductive microporous carbon framework with excellent electrical and ionic properties, which not only enables effective dynamics, but also improves the mechanical strength of the zinc anode. The cycling performance of Zn-POME has improved at 100% DOD [52].

In addition to coating nano-sized particles, ion exchange coatings can also be applied to zinc webs. In the process of commercialization, zinc mesh is easier to achieve large-scale production. The passivation and dissolution of zinc anodes can also be solved by applying graphene oxide (GO) to the surface of the zinc mesh.

The study combines the selective light transmittance of ion-exchange coatings to prevent dissolution with the electrical conductivity of cohesive coatings to prevent passivation. GO coated zinc anodes remain intact after the cycle (DOD ~ 15%), which can be inferred that the GO coating does not break due to volume expansion from zinc to zinc oxide. Ionomer hydroxide-conductive polymer (IHCP) can also be used to coat anodic particles. The zinc oxide layer is deposited on a carbon mesh (ZnO/C) and coated with IHCP. IHCP can achieve selective penetration of OH^- and $Zn(OH)_4^{2-}$, because it has a part of cationic head groups covalently bonded to the carbon net, so it can penetrate to the zinc anode. The improved anode achieves higher specific discharge capacity and good cycle stability at 53.6% DOD. In particular, the hardness and toughness of IHCP can be successfully adapted to the volume variation of the active material during circulation [51].

4.2.3 Inhibition of Dendrite Formation by Membranes and Additives

The damage of dendritic growth to cells can be mitigated by physically blocking the contact between electrolyte and electrode or changing the electrochemical deposition behavior of zinc in the electrolyte. At present, researchers mainly use polymer separators and introduce various additives into the battery system to reduce the dendrite problem. Polymer separators can inhibit the penetration caused by dendritic growth by establishing a physical barrier. Polymer separators used in zinc-air batteries mainly include PVA film, PE film or hydrated cellulose film [53]. In addition to the use of commercial separators such as Celgard 5550, Nafion 115, and polyvinyl alcohol, various new preparation methods with high anionic selectivity and low ionic resistance have been developed [54]. However, the polymer separator could not prevent the formation and growth of zinc dendrites fundamentally. Therefore, it is better to use it in conjunction with other methods.

The dendrites of zinc can be inhibited by changing the electrochemical deposition behavior of zinc in the electrolyte by introducing organic or inorganic additives into the electrode and electrolyte. More recently, some organic agents have been added to the electrolyte to produce flat zinc-based films composed of small grains by influencing the electrochemical performance of zinc electrodes during electrodeposition, such as grain size, chemical composition, and corrosion resistance [58–60]. Some studies have confirmed that zinc dendritic deposition and zinc electrode corrosion during charge and discharge measurements of zinc-air batteries can be inhibited by organic additives such as EDTA, Tween 20, and tartaric acid [60, 61]. According to the SEM image in Fig. 4.5b, the prevention of dendritic crystallization follows [55]. In addition, a variety of organic additives can be introduced in the form of polymer adhesives to prepare composite electrodes. For example, ZnO/PVA/β-Cd /PEG composite electrodes could be obtained by embedding PEG into β-cyclodextrin (β-CD) and cross-linking with PVA via glutaraldehyde. The polymer binder could form a fixed electrode with a stable three-dimensional network structure, which tightly wrapped the anode active material, thus reducing the deformation of the zinc anode and improving the cycling stability of the composite electrode [62].

Despite their many advantages, organic additives can act as insulators and impurities in electrolytes and electrodes, increasing battery impedance and weakening battery performance. Therefore, inorganic additives, such as CdO, PbO, Pb_3O_4, Bi_2O_3, In_2O_3, and SnO, have also been studied to mitigate the growth of zinc dendrites [63–66]. The dissolution of these metal oxide additives in the electrolyte results in the formation of metal ions at a higher reduction potential than metal zinc. Therefore, in the process of electrochemical deposition, the metal elements in the additive can easily combine with zinc and affect the growth of zinc dendrite. However, the potential hazards of metal-containing materials such as Pb exist, and the exact mechanism of additive influence on the cathode efficiency and morphology of zinc electrodeposition remains to be discussed [67]. In addition, adding an appropriate amount of inorganic additives in the electrode preparation process can control the overpotential to uniformly reduce zinc, as shown in Fig. 4.5c [56].

Compared with other organic additives, room temperature ionic liquids (RTILs) have attracted great attention in recent years due to their wide electrochemical window, high safety, and good solubility. 1-ethyl-3-methylimidazolium (EMI-based) ionic liquids have been studied for the inhibition of zinc dendrites [68–70]. In particular, 1-ethyl-3-methyl-imidazolium dicyanamide (EMI-DCA) can effectively prevent the dendritic growth of zinc by changing the morphology of zinc deposits, because the zinc reduction process and the zinc film/electrolyte interface are changed by ionic liquid during charging to inhibit uneven deposition of zinc.

4.2.4 Other Methods to Inhibit Dendrite Formation

With the rise of personalized portable and flexible electronic products, solid electrolytes have attracted people's attention [71]. As a polymer gel electrolyte, sodium polyacrylate hydrogel (PANa) electrolyte has an inhibitory effect on dendritic material because acrylate ions promote the formation of quasi-solid electrolyte interfaces [57]. As shown in Fig. 4.5d, electrostatic interaction between negatively charged acrylate groups and positively charged zinc ions along the PANa electrolyte chain results in the formation of a quasi-solid electrolyte interface (quasi-SEI) without dendrimers.

Changing the battery charging pattern is also a way to curb the growth of the twigs. Concentration polarization is easily induced by a constant current charging mode to promote dendritic growth. On the contrary, the zincate can be diffused to the negative electrode under the electric field with the reverse voltage generated by the pulse charge, reducing the concentration polarization caused by the charge, and some branches can also be dissolved, making the zinc electrode surface smooth [72]. Therefore, pulse charging mode can effectively reduce concentration polarization and inhibit dendritic growth.

4.2.5 Effect of Corrosion Inhibitor on Self-discharge

The main method to inhibit HER reaction and reduce the corrosion rate is to add inorganic or organic corrosion inhibitors into the electrode or electrolyte to enhance the evolution potential of anode hydrogen or to form a surface adsorption layer. Inorganic corrosion inhibitors mainly include metal elements, metal oxides, metal hydroxides, and salts. Through the use of alloys, coatings and additives, HER can be inhibited on metals with high chemical stability and high hydrogen evolution overpotential, such as lead, cadmium, indium, mercury, bismuth, nickel, etc. [73–76]. For example, zinc alloy electrodes with good corrosion resistance can be formed by introducing different amounts of metal Bi. Since the prepared material has a large hydrogen overpotential, the corrosion reaction is indeed inhibited by mechanical alloying [77]. The synthetic shell prevents zinc particles from interacting directly with the alkaline electrolyte, which is another way to block HER. This method usually involves coating zinc particles with other metal oxides, such as alumina (Al_2O_3), bismuth oxide (Bi_2O_3), and indium oxide (In_2O_3) [75]. Among them, aluminum oxide-coated zinc anode (Fig. 4.6b) has the smallest amount of hydrogen evolution and has been shown to have the most positive effect on inhibiting HER reaction (Fig. 4.6a) [78]. However, it is important to note that metal or metal compound additives may have higher toxicity and negative environmental effects. Also, some metals are too expensive to be used on a large scale, such as In [77]. Currently, silica coatings have been chosen because of their ability to form hydroxides such as $Si(OH)_4$ that prevent the hydroxide from adsorbing to the metal zinc and thus preventing zinc corrosion [79].

More and more people pay attention to safety, pollution-free organic corrosion inhibitors, such as PEG, FPEA, and IMZ [80, 81]. Among them, polyethylene glycol (PEG) is the simplest compound containing polyoxyethylene, with good hydrophilicity, acid and alkali resistance, and high stability, is one of the most common electrolyte organic corrosion inhibitors, the inhibition efficiency of compound inhibitor is better than that of single inhibitor. Inhibitors such as Tween 20 and IMZ can be combined with PEG to exert a synergistic effect on zinc corrosion. The compound inhibitor can be completely adsorbed on the active site of zinc surface and achieve better inhibition efficiency than the single inhibitor [82]. Some conductive polymers, such as polyaniline (PANI), are considered the most promising zinc anodic coatings because of their extensive electrical conductivity, high stability, adjustable properties, low cost, and ease of synthesis [83, 84]. When polyaniline is between zinc active material and water electrolyte in the zinc-air battery, its good shielding effect can inhibit corrosion reaction and HER [81]. As is displayed in Fig. 4.6c, 20PANI@Zn showed an 85% corrosion inhibition efficiency for pure zinc, with a capacity retention rate of 97.81% after 24 h storage at ambient temperature compared to no storage. However, the capacity of zinc anodes containing PANI is lower than that of pure zinc anodes before 24 h, so subsequent studies need to pay attention to the negative effect of corrosion inhibitors on battery performance before storage.

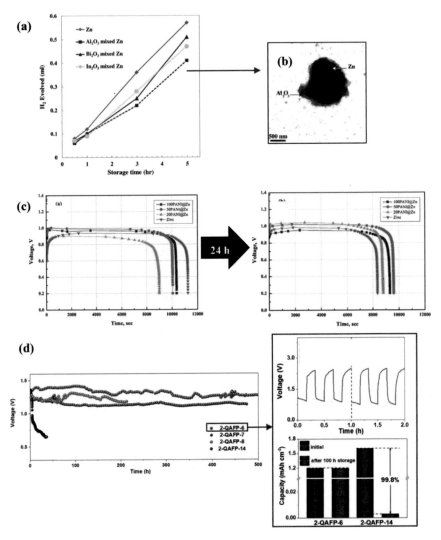

Fig. 4.6 **a** At 60 °C, the volume of hydrogen spontaneous evaporation on the surface of zinc gel anode was measured [75]. **b** Transmission electron microscope image of zinc particles coated with Al₂O₃ [77]. **c** PANI@Zn electrochemical potential curves at ambient temperature for 24 h [85]. **d** OCV of QAFP electrolyte based batteries about time, charging capacity, and capacity before and after storage [86]

4.2.6   Long Shelf Life Based on Solid Electrolyte

Currently, alkali-solid polymer electrolytes are used in flexible solid Zn-air batteries [87]. Due to the spontaneous reaction between zinc and alkaline solid electrolyte, self-discharge occurs during charge and discharge reactions, even at rest [76]. In addition, it should be noted that shelf life is threatened by electrolyte failures, such as electrolyte evaporation and carbonization. In particular, KOH reacts with CO_2 in the atmosphere to form carbonate, the electrolyte may dry out into the air, and then the battery won't work without enough OH^-. Solid electrolytes with good water retention properties were developed [88]. In solid-state electrolytes, near-neutral components are usually used instead of alkaline electrolytes to inhibit self-discharge by preventing spontaneous HER reaction of metal zinc. Solid electrolytes, such as electrolyte anion exchange membranes (AEM), can construct nanochannels that allow rapid ion transport directionally, preserving high ionic conductivity for near-neutral components [89]. As seen in Fig. 4.6d, batteries using a near-neutral quaternary ammonium functionalized polyvinyl alcohol (QAFP) electrolyte (pH = 6) showed the best stability, without any OCV loss over 500 h [86]. After a long period of storage, the battery can remain rechargeable and its self-discharge rate is only 7% per month.

References

1. J.S. Lee, S.T. Kim, R. Cao, N.S. Choi, M. Liu, K.T. Lee, J. Cho, Adv. Energy Mater. **1**, 34 (2011)
2. H.-F. Wang, C. Tang, Q. Zhang, Adv. Funct. Mater. **28**, 1803329 (2018)
3. Y. Shao, J.P. Dodelet, G. Wu, P. Zelenay, Adv. Mater. **31**, 1807615 (2019)
4. J. Zhang, Q. Zhou, Y. Tang, L. Zhang, Y. Li, Chem. Sci. **10**, 8924 (2019)
5. J. Yi, P. Liang, X. Liu, K. Wu, Y. Liu, Y. Wang, Y. Xia, J. Zhang, Energy Environ. Sci. **11**, 3075 (2018)
6. J. Fu, Z.P. Cano, M.G. Park, A. Yu, M. Fowler, Z. Chen, Adv. Mater. **29**, 1604685 (2017)
7. M. Qiao, C. Tang, L.C. Tanase, C.M. Teodorescu, C. Chen, Q. Zhang, M.M. Titirici, Mater. Horiz. **4**, 895 (2017)
8. Y. Li, H. Dai, Chem. Soc. Rev. **43**, 5257 (2014)
9. V. Caramia, B. Bozzini, Mater. Renew. Sustain. Energy **3**, 28 (2014)
10. Z. Zhou, Y. Zhang, P. Chen, Y. Wu, H. Yang, H. Ding, Y. Zhang, Z. Wang, X. Du, N. Liu, Chem. Eng. Sci. **194**, 142 (2019)
11. L. Zhang, X. Yang, R. Cai, C. Chen, Y. Xia, H. Zhang, D. Yang, X. Yao, Nanoscale **11**, 826 (2019)
12. J. Pan, Y.Y. Xu, H. Yang, Z. Dong, H. Liu, B.Y. Xia, Adv. Sci. **5**, 1700691 (2018)
13. J. Pan, X.L. Tian, S. Zaman, Z. Dong, H. Liu, H.S. Park, B.Y. Xia, Batteries Supercaps **2**, 336 (2019)
14. Y. Guo, Y.N. Chen, H. Cui, Z. Zhou, Chin. J. Catal. **40**, 1298 (2019)
15. W. Xiao, M.A.L. Cordeiro, G. Gao, A. Zheng, J. Wang, W. Lei, M. Gong, R. Lin, E. Stavitski, H.L. Xin, D. Wang, Nano Energy **50**, 70 (2018)
16. A.A. Mohamad, J. Power Sources **159**, 752 (2006)
17. D.U. Lee, J.-Y. Choi, K. Feng, H.W. Park, Z. Chen, Adv. Energy Mater. **4**, 1301389 (2014)
18. E.D. Farmer, A.H. Webb, J. Appl. Electrochem. **2**, 123 (1972)
19. K. Wang, P. Pei, Z. Ma, H. Chen, H. Xu, D. Chen, X. Wang, J. Mater. Chem. A **3**, 22648 (2015)
20. C. Chakkaravarthy, H.V.K. Udupa, J. Power Sources **10**, 197 (1983)
21. M. Bockelmann, M. Becker, L. Reining, U. Kunz, T. Turek, J. Electrochem. Soc. **166**, A1132 (2019)
22. M. Bockelmann, M. Becker, L. Reining, U. Kunz, T. Turek, J. Electrochem. Soc. **165**, A3048 (2018)
23. Y. Wu, Y. Zhang, Y. Ma, J.D. Howe, H. Yang, P. Chen, S. Aluri, N. Liu, Adv. Energy Mater. **8**, 1802470 (2018)

24. J.F. Parker, C.N. Chervin, E.S. Nelson, D.R. Rolison, J.W. Long, Energy Environ. Sci. **7**, 1117 (2014)
25. J.F. Parker, C.N. Chervin, I.R. Pala, M. Machler, M.F. Burz, J.W. Long, D.R. Rolison, Science **356**, 415 (2017)
26. M.B. Liu, G.M. Cook, N.P. Yao, J. Electrochem. Soc. **128**, 1663 (1981)
27. S. Thomas, I.S. Cole, M. Sridhar, N. Birbilis, Electrochim. Acta **97**, 192 (2013)
28. J.W. Diggle, A.R. Despic, J.O.M. Bockris, J. Electrochem. Soc. **116**, 1503 (1969)
29. A.R. Despic, J. Diggle, J.O.M. Bockris, J. Electrochem. Soc. **115**, 507 (1968)
30. W. Lu, C. Xie, H. Zhang, X. Li, Chemsuschem **11**, 3996 (2018)
31. M. Biton, F. Tariq, V. Yufit, Z. Chen, N. Brandon, Acta Mater. **141**, 39 (2017)
32. B.S. Lee, S. Cui, X. Xing, H. Liu, X. Yue, V. Petrova, H.D. Lim, R. Chen, P. Liu, A.C.S. Appl, Mater. Interfaces **10**, 38928 (2018)
33. R.E.F. Einerhand, W.H.M. Visscher, E. Barendrecht, J. Appl. Electrochem. **18**, 799 (1988)
34. K.M.S. Youssef, C.C. Koch, P.S. Fedkiw, Corros. Sci. **46**, 51 (2004)
35. H. Yang, Y. Cao, X. Ai, L. Xiao, J. Power Sources **128**, 97 (2004)
36. K. Liu, P. He, H. Bai, J. Chen, F. Dong, S. Wang, M. He, S. Yuan, Mater. Chem. Phys. **199**, 73 (2017)
37. Z. Hou, X. Zhang, X. Li, Y. Zhu, J. Liang, Y. Qian, J. Mater. Chem. A **5**, 730 (2017)
38. Y. Qiang, S. Zhang, L. Guo, S. Xu, L. Feng, I.B. Obot, S. Chen, J. Cleaner Prod. **152**, 17 (2017)
39. M.A. Deyab, J. Power Sources **292**, 66 (2015)
40. R.K. Ghavami, Z. Rafiei, S.M. Tabatabaei, **164**, 934 (2007)
41. R. Khayat Ghavami, F. Kameli, A. Shirojan, A. Azizi, *J. Energy Storage*, **7**, 121 (2016)
42. S. Hosseini, W. Lao-atiman, S.J. Han, A. Arpornwichanop, T. Yonezawa, S. Kheawhom, Sci. Rep. **8**, 14909 (2018)
43. M.C.H. McKubre, D.D. Macdonald, J. Electrochem. Soc. **128**, 524 (1981)
44. H.E. Lin, C.H. Ho, C.Y. Lee, Surf. Coat. Technol. **319**, 378 (2017)
45. M.L. Green, R.A. Levy, JOM **37**, 63 (1985)
46. K.C. Huang, K.S. Chou, Electrochem. Commun. **2007**, 9 (1907)
47. J.F. Parker, E.S. Nelson, M.D. Wattendorf, C.N. Chervin, J.W. Long, D.R. Rolison, A.C.S. Appl, Mater. Interfaces **6**, 19471 (2014)
48. S.W. Bian, I.A. Mudunkotuwa, T. Rupasinghe, V.H. Grassian, Langmuir **27**, 6059 (2011)
49. Y. Zhang, Y. Wu, H. Ding, Y. Yan, Z. Zhou, Y. Ding, N. Liu, Nano Energy **53**, 666 (2018)
50. M. Schmid, M. Willert-Porada, J. Power Sources **351**, 115 (2017)
51. D. Stock, S. Dongmo, D. Damtew, M. Stumpp, A. Konovalova, D. Henkensmeier, D. Schlettwein, D. Schröder, A.C.S. Appl, Energy Mater. **1**, 5579 (2018)
52. P. Chen, Y. Wu, Y. Zhang, T.H. Wu, Y. Ma, C. Pelkowski, H. Yang, Y. Zhang, X. Hu, N. Liu, J. Mater. Chem. A **6**, 21933 (2018)
53. C.-c. Yang, J. Ming, C.-y. Wu, **191**, 669 (2009)
54. P. Kritzer, J.A. Cook, J. Electrochem. Soc. **154**, A481 (2007)
55. M.C. Huang, S.H. Huang, S.C. Chiu, K.L. Hsueh, W.S. Chang, C.C. Yang, C.C. Wu, J.C. Lin, J. Chin. Chem. Soc. **65**, 1239 (2018)
56. D.J. Park, E.O. Aremu, K.S. Ryu, Appl. Surf. Sci. **456**, 507 (2018)
57. Y. Huang, Z. Li, Z. Pei, Z. Liu, H. Li, M. Zhu, J. Fan, Q. Dai, M. Zhang, L. Dai, C. Zhi, Adv. Energy Mater. **8**, 1802288 (2018)
58. O. Aaboubi, J. Douglade, X. Abenaqui, R. Boumedmed, J. Vonhoff, Electrochim. Acta **56**, 7885 (2011)
59. M.R.H.D. Almeida, E.P. Barbano, M.F.D. Carvalho, I.A. Carlos, J.L.P. Siqueira, L.L. Barbosa, Surf. Coat. Technol. **206**, 95 (2011)
60. H. Zhou, Q. Huang, M. Liang, D. Lv, M. Xu, Mater. Chem. Phys. **128**, 214 (2011)
61. J. Torrent-Burgués, E. Guaus, J. Appl. Electrochem. **37**, 643 (2007)
62. Z. Zhang, D. Zhou, G. Huang, L. Zhou, B. Huang, J. Electroanal. Chem. **827**, 85 (2018)
63. J.W. Diggle, A. Damjanovic, J. Electrochem. Soc. **119**, 1649 (1972)
64. J.M. Wang, L. Zhang, C. Zhang and J. Q. Zhang **102**, 139 (2001)
65. R.Y. Wang, D.W. Kirk, G.X. Zhang, J. Electrochem. Soc. **153**, C357 (2006)

66. Y.-h. Wen, J. Cheng, L. Zhang, X. Yan, Y.-s. Yang, **193**, 890 (2009)
67. H.I. Kim, H.C. Shin, J. Alloys Compd. **645**, 7 (2015)
68. Y. Song, J. Hu, J. Tang, W. Gu, L. He, X. Ji, A.C.S. Appl, Mater. Interfaces **8**, 32031 (2016)
69. M. Xu, D.G. Ivey, W. Qu and Z. Xie **274**, 1249 (2015)
70. T.J. Simons, D.R. Macfarlane, M. Forsyth, P.C. Howlett, ChemElectroChem **1**, 1688 (2014)
71. Y. Li, J. Fu, C. Zhong, T. Wu, Z. Chen, W. Hu, K. Amine, J. Lu, Adv. Energy Mater. **9**, 1802605 (2019)
72. G. Garcia, E. Ventosa, W. Schuhmann, A.C.S. Appl, Mater. Interfaces **9**, 18691 (2017)
73. C. Woo, K. Sathiyanarayanan, S. Wook, H. Soo and M. Soo **159**, 1474 (2006)
74. H.S. Kim, Y.N. Jo, W.J. Lee, K.J. Kim, C.W. Lee, Electroanalysis **27**, 517 (2015)
75. S.M. Lee, Y.J. Kim, S.W. Eom, N.S. Choi, K.W. Kim, S.B. Cho, J. Power Sources **227**, 177 (2013)
76. M. Yano, S. Fujitani, K. Nishio, Y. Akai, M. Kurimura, J. Power Sources **74**, 129 (1998)
77. Y.N. Jo, K. Prasanna, S.H. Kang, P.R. Ilango, H.S. Kim, S.W. Eom, C.W. Lee, J. Ind. Eng. Chem. **53**, 247 (2017)
78. K. Wongrujipairoj, L. Poolnapol, A. Arpornwichanop, S. Suren, S. Kheawhom, *Physica Status Solidi (b)*, **254**, 1600442 (2017)
79. M. Schmid, U. Schadeck, Surf. Coat. Technol. **310**, 51 (2017)
80. J. Dobryszycki, S. Biallozor, Corros. Sci. **43**, 1309 (2001)
81. M. Naja, N. Penazzi, G. Farnia, G. Sandonà, Electrochim. Acta **38**, 1453 (1993)
82. M. Liang, H. Zhou, Q. Huang, S. Hu, W. Li, J. Appl. Electrochem. **41**, 991 (2011)
83. D. Zhang, Y. Wang, Mater. Sci. Eng., B **134**, 9 (2006)
84. S. Bhadra, D. Khastgir, N.K. Singha, J. Hee, **34**, 783 (2009)
85. Y. Nam, S. Hyun, K. Prasanna, S. Wook, C. Woo, Appl. Surf. Sci. **422**, 406 (2017)
86. C. Lin, S.S. Shinde, X. Li, D.H. Kim, N. Li, Y. Sun, X. Song, H. Zhang, C.H. Lee, S.U. Lee, J.H. Lee, Chemsuschem **11**, 3215 (2018)
87. X. Fan, J. Liu, Z. Song, X. Han, Y. Deng, C. Zhong, W. Hu, Nano Energy **56**, 454 (2019)
88. M. Li, B. Liu, X. Fan, X. Liu, J. Liu, J. Ding, X. Han, Y. Deng, W. Hu, C. Zhong, A.C.S. Appl, Mater. Interfaces **11**, 28909 (2019)
89. J. Fu, J. Zhang, X. Song, H. Zarrin, X. Tian, J. Qiao, L. Rasen, K. Li, Z. Chen, Energy Environ. Sci. **9**, 663 (2016)

Chapter 5
Electrolyte of Zn-Air Battery

A conventional Zn-air battery consists of a zinc electrode, an electrolyte, and an air electrode. Electrolyte, as the "blood" of zinc empty battery, plays a crucial role. Electrolytes are divided into solid electrolytes and liquid electrolytes. Whether solid or liquid, the electrolyte system of zinc empty battery is mostly potassium hydroxide. In this chapter, the research progress of different systems of the Zn-air battery was briefly introduced from the perspective of the Zn-air battery electrolyte, and the advantages and disadvantages of the Zn-air battery were discussed.

5.1 Liquid Electrolyte

5.1.1 Alkaline Electrolyte

The Zn-air battery has used alkaline electrolytes, such as KOH, NaOH, and other liquid electrolytes. Among them, KOH has better ionic conductivity, a higher oxygen diffusion rate, and lower viscosity than NaOH. On the one hand, KOH has a lower viscosity and a higher diffusion coefficient of oxygen, because the high concentration of KOH solution ensures higher ionic conductivity and prevents H_2 on the Zn surface. Under these conditions, the diffusivity and solubility of oxygen are reduced, which allows gaseous oxygen to be used more efficiently. On the other hand, KOH exposed to air can be associated with CO_2. The reaction produces $KHCO_3$ with greater solubility or K_2CO_3. In this way, the porosity of the cathode gas diffusion layer is blocked by carbonate precipitation and the performance of the battery is weakened. The commercial electrolyte was 0.2 M $Zn(Ac)_2$ and 6 M KOH mixed solution. The addition of zinc salts can greatly improve the rechargeable characteristics of the battery.

© The Author(s), under exclusive license to Springer Nature Singapore Pte Ltd. 2023   175
S. Peng, *Zinc-Air Batteries*, https://doi.org/10.1007/978-981-19-8214-9_5

5.1.1.1 Water Loss of the Electrolyte

The Zn-air battery typically consists of a zinc anode, an alkaline electrolyte, a separator, and an air cathode (a carbon layer containing a catalyst). During the discharge, oxygen from the atmosphere diffuses to the cathode and is reduced to hydroxyl ions. These hydroxyl ions migrate to the anode, where they react with zinc to form $Zn(OH)_4{}^{2-}$ and release electrons. These electrons are transported to the cathode, and the $Zn(OH)_4{}^{2-}$ then precipitates to form zinc oxide. The anode, cathode, and total reaction of the Zn-air battery in the discharge process can be expressed as:

Anode:

$$Zn + 4OH^- \rightarrow 4Zn(OH)_4^{2-} + 2e^-$$

$$Zn(OH)_4^{2-} \rightarrow ZnO + H_2O + 2OH^-$$

Cathode:

$$O_2 + 2H_2O + 4e^- \rightarrow 4OH^-$$

The total response:

$$2Zn + O_2 \rightarrow 2ZnO$$

The cathode can consume water from $Zn(OH)_4{}^{2-}$. The total reaction product is zinc oxide. However, the Zn-air battery is an open system, in the process of long-term work in the air, due to the difference between the battery system and the environment in temperature, water vapor partial pressure, and other aspects, the battery electrolyte will sometimes absorb water from the air or water evaporation. These phenomena are related to the concentration of electrolytes, relative humidity of the air, ambient temperature, the structure of the cathode, and the flow rate of air. If 28%, 30%, and 32% KOH are used as electrolytes respectively, the saturated vapor pressure of the Zn-air battery is about 64%, 60%, and 55% of relative humidity at 25 °C. When the relative humidity of the atmosphere is lower than the above value, the battery will lose water, which causes the battery to dry up gradually and the battery performance deteriorates. When the relative humidity of the atmosphere is higher than the corresponding value, the battery will gradually absorb the moisture in the atmosphere, so that the concentration of the battery electrolyte gradually thins. In several cases, the cathode will be covered by water molecules, and the electrochemical reaction sites will be reduced, resulting in the degradation of battery performance.

In aqueous electrolytes, acidic electrolytes including inorganic acids such as methanesulfonic acid, polyvinyl sulfonic acid, and polyvinyl sulphuric acid have been proposed for Zn-air batteries. However, in most acidic electrolytes, the zinc metal in the anode is unstable and presents serious corrosion problems. Therefore, acid electrolytes should not be used in practical applications. As for alkaline aqueous electrolytes widely used, there are still more or fewer problems in the Zn-air battery system. Therefore, much research is focused on developing new electrolyte systems to modify or replace traditional alkaline electrolytes.

5.1.1.2 Passivation

Passivation refers to the phenomenon that ZnO supersaturated in KOH solution precipitates and nucleates on the surface of the zinc electrode, thus blocking the electrode surface and affecting the further reaction of the electrode. The causes of passivation include electrolyte aging, current density, and zincate stratification tendency [1, 2]. There is relatively little research on the passivation and there is no unified understanding of the mechanism of passivation.

Hampson et al. believed that passivation occurred when ZnO was formed on the zinc electrode surface [3–5]. The passivation time varies from 10 to 800 s with the current density and the concentration of KOH solution. L. Liu et al. considered that ZnO loosely attached to the surface of zinc electrode did not affect OH^- at the low current density. The transfer of ions did not cause electrode passivation. Passivation occurs only when a dense ZnO layer is formed on the surface of the zinc electrode at a high current density.

Thomas et al. [6] studied the composition of the passivation layer of zinc under alkaline conditions by focusing on an ion beam scanning electron microscope. It is found that the passivation layer consists of two parts. The inner layer is a dense oxide layer, and the outer layer is mainly $ZnO/Zn(OH)_2$ sediments. When pH $= 13$, the outer layer is loose and unstable, which increases the corrosion degree of zinc. At pH $= 12$, the outer layer is very stable, preventing the electrolyte from contacting the zinc.

Powers et al. [7] found that the convection condition of electrolytes had a great influence on the physical and chemical properties of the zinc passivation film. When there is no convection in the electrolyte, the white loose flocculent type I film is formed near the surface of the zinc electrode, which comes from the precipitation and deposition of saturation zincate. When the electrolyte is stirred, light gray or even black dense type II film is formed on the surface of the zinc electrode, which is directly formed on the electrode surface. The authors believe that type II film is the passivation film that leads to decreased electrode activity.

Yang et al. [8] studied the effect of sodium dodecylbenzene sulfonate (SDBS) in dilute alkali solution on zinc passivation. It is found that SDBS can be adsorbed on the surface of zinc electrode so that only loose porous passive film can be formed on the surface of zinc electrode. The passivation film can modify the electrode surface, which not only delays the passivation of zinc electrode to improve the utilization rate of the electrode but also promotes the diffusion of electrode discharge products. Compared with that without SDBS, the discharge capacity of a zinc electrode in the experimental group with SDBS was increased by 35%.

5.1.1.3   Hydrogen Evolution Corrosion

Metallic zinc is thermodynamically unstable in alkaline solution and will undergo hydrogen evolution corrosion. The hydrogen produced by the corrosion reaction can raise the pressure inside the battery, causing it to bulge and even damage it. The corrosion reaction consumes zinc, an electrochemically active substance, which also reduces battery capacity. The conjugated corrosion reaction of hydrogen evolution is shown in Eqs. (5.1) and (5.2).

$$Zn + 4OH^- \rightarrow 4Zn(OH)_4^{2-} + 2e^- \tag{5.1}$$

$$2H_2O + 2e^- \rightarrow 2OH^- + H_2 \tag{5.2}$$

To inhibit the zinc corrosion reaction and reduce the corrosion rate, the main solution at present is to add a corrosion inhibitor to the electrode or electrolyte. The corrosion inhibitor includes inorganic corrosion inhibitor, organic corrosion inhibitor, and compound corrosion inhibitor. The inorganic corrosion inhibitors mainly achieve corrosion inhibition by increasing the hydrogen evolution overpotential of zinc electrodes. The organic corrosion inhibitor achieves the effect of corrosion inhibition by forming a physical barrier on the electrode surface through physical or chemical adsorption. Organic and inorganic corrosion both have advantages and disadvantages, so organic–inorganic composite corrosion inhibitors are getting more and more attention.

Inorganic corrosion inhibitors mainly include metal elements, metal oxides, metal hydroxides, and their salts. A protective film was formed on the surface of zinc by alloying or substitution reaction to improve the hydrogen evolution overpotential of the zinc electrode and achieve the purpose of corrosion inhibition. Indium metal has low resistivity, good softness, strong chemical stability, high hydrogen evolution overpotential, and good affinity with zinc. The contact resistance between particles can be reduced by doping indium into zinc particles. So, indium is the most widely used corrosion inhibitor of mercury substitution. In addition, the metals that are widely concerned include mercury, lead, cadmium, bismuth, tin, aluminum, calcium, barium, and gallium. Mercury, lead, and cadmium have high hydrogen evolution overpotential to achieve a good corrosion inhibition effect. But its usage is restricted because of its toxicity, which causes harm to both the environment and the human body.

Kannan et al. [9] studied the corrosion behavior of zinc electrodes in 10 M NaOH solution by electrolyte additives (sodium citrate, sodium stannate, calcium oxide) and metal alloys (magnesium, aluminum, lead). In the absence of additives, the corrosion process is controlled by Eq. (5.2). When additives such as sodium citrate, sodium stannate, or calcium oxide are added to the electrolyte, the corrosion process is controlled by Eq. (5.1). When the number of additives is 15% sodium citrate + 0.3% calcium oxide is the best corrosion inhibition effect. The results of the study on the alloy showed that the temperature range was 25 ~ 120 mA cm^{-2}. The alloy composed of Zn + 0.01%, Mg + 0.01% Pb has the highest anode efficiency in the range of current density. This indicates that the magnesium-lead alloy is also a kind of potential mercury substitute.

Hg, Cr, and other inorganic zinc inhibitors have a good corrosion inhibition effect, but their use is limited because of their toxicity. The corrosion inhibition effect of other inorganic corrosion inhibitors is not ideal, so a wide variety of organic zinc corrosion inhibitors, which are safe and pollution-free, are attracting more and more attention. With the development of the research on organic zinc corrosion inhibitors, the relationship between the structure and dielectric properties of organic compounds and corrosion inhibition performance has been gradually recognized. According to the adsorption theory, an organic corrosion inhibitor forms a physical barrier layer on the electrode surface through physical or chemical adsorption, to achieve corrosion inhibition. According to electrochemical theory, organic corrosion inhibitor adsorbed on the electrode surface occupies the active site of corrosion reaction on the electrode surface. This increases the activation energy of cathodic or anodic reactions, thus achieving corrosion inhibition. Heterocyclic compounds and surfactants are commonly used as organic corrosion inhibitors. For heterocyclic zinc inhibitors, N, O, P, S, and other heterocyclic atoms or π electron systems are often needed in the structure, so that it is easier to adsorb on the surface of zinc. The stronger the adsorption, the better the inhibition effect. Surfactant zinc corrosion inhibitor mainly achieves corrosion inhibition through the action of hydrophilic and hydrophobic groups. The hydrophilic end is adsorbed on the surface of the zinc. The hydrophobic end is to reduce the concentration of H_2O and OH^- near the hydrophobic layer to achieve the purpose of corrosion. The more stable the hydrophilic end is combined with zinc means that the stronger the hydrophobicity formed by the hydrophobic end, the better the effect of the corrosion.

The hydrophilicity of polyoxyethylene groups is good and has high acid resistance and high stability. Polyethylene glycol (PEG), the simplest compound containing polyoxyethylene, has been extensively studied. Dobryszycki et al. studied the zinc corrosion inhibition effect of PEG with different molecular weights in 7 M KOH solution [10], and compared it with commercialized surfactant, and FPEA [11], which had a better corrosion inhibition effect than that of reported works by other researchers. The polarization curves show that PEG and FPEA can inhibit corrosion by inhibiting the cathodic branch of the corrosion reaction. PEG400 has better corrosion inhibition performance than PEG200 and PEG600. PEG400 and FPEA are effective zinc corrosion inhibitors. It can be inferred from the electron microscope photos of the zinc anode after the corrosion experiment that PEG400 can strongly adsorb on the surface of the zinc anode to prevent zinc corrosion. Based on the analysis of the hydrogen evolution volume-time curve, the author concluded that the corrosion of zinc in the alkaline condition is electrochemical corrosion followed by chemical corrosion.

The inorganic corrosion inhibitor can improve the hydrogen evolution overpotential and the deposition morphology of zinc, but its inhibition on the anode branch of corrosion reaction is far less than that of an organic corrosion inhibitor. Organic corrosion inhibitors can not only improve the overpotential of hydrogen evolution but also inhibit anodic dissolution. Therefore, to make full use of the respective characteristics of inorganic and organic corrosion inhibitors and play their synergistic role, the two types of zinc corrosion inhibitors are often added together in practical application.

Zhe et al. studied composite corrosion inhibitor [PEG600 + In(OH)$_3$] corrosion inhibition performance of zinc in alkaline conditions [12, 13]. The electrolyte was a 10 mol L^{-1} KOH solution containing 0.86 M ZnO. The results showed that PEG600 and In(OH)$_3$ had an obvious synergistic corrosion inhibition effect. On the one hand, the indium generated by electrode surface replacement strengthens the adsorption strength of PEG600. On the other hand, PEG600 is adsorbed on the surface of zinc, which inhibits the precipitation of hydrogen cathode near the open circuit potential, and the anodic dissolution of zinc. However, when the zinc electrode anode is polarized to PEG600 desorption potential, PEG600 desorption from the electrode surface no longer inhibits the anodic dissolution of zinc.

S. N. Hong et al. studied lead nitrate (Pb(NO$_3$)$_2$), sodium dodecylbenzene sulfonate (SDBS), and their composite corrosion inhibitors on the corrosion behavior of zinc in 3 mol L^{-1} KOH solution. The results showed that the hydrogen evolution of the zinc anode was serious without additives, and was inhibited only with the addition of SDBS. Meanwhile, after adding Pb(NO$_3$)$_2$ (10 mg L^{-1}) and SDBS (500 mg L^{-1}), zinc negative hydrogen analysis is the highest, and the corrosion efficiency reaches 81.1%. Lead nitrate can replace zinc and cover the surface of the zinc electrode, and lead contains p vacant orbital which can accept lone pair electrons of SDBS in chain structure and adsorb it on the surface of the electrode. In this way, the advantages of high hydrogen evolution overpotential of Pb and the barrier function of SDBS can be simultaneously played, to achieve a better corrosion inhibition effect.

5.1.1.4 Zinc Dendrite

In an alkaline Zn-air battery, dendritic zinc crystals will be generated in the negative electrode when charging, which is called zinc dendrite. The charging reaction process of the zinc electrode is mainly controlled by the liquid mass transfer process, and the concentration of reactive substances near the surface of the zinc electrode is very low, resulting in a large concentration polarization. The reactive substances in the electrolyte body are more likely to diffuse to the electrode surface, the current distribution on the electrode is uneven, and finally, the dendrite is formed. At the initial stage of dendrite growth, the dendrite length increases exponentially with time, after which the total current is linear with the square of time. On the one hand, zinc dendrites can pierce the battery diaphragm and cause battery short circuit failure. At the same time, it will fall off the electrode surface, resulting in battery capacity attenuation, resulting in a shortened battery life.

J. M. Wang et al. added Bi^{3+} and tetrabutylammonium bromide (TBAB) to the electrolyte to study the growth behavior of zinc dendrite in alkaline zincate solution [14]. Bismuth ions preferentially deposited on the surface of zinc electrode during cathodic polarization, which inhibited the formation of zinc dendrites and hardly affected the anodic dissolution behavior of zinc. TBAB can effectively inhibit the generation of zinc dendrites in the low cathode polarization region by adsorption on the electrode surface, but cannot inhibit the generation of zinc dendrites in the high cathode polarization region due to the desorption of TBAB. Bismuth ion and TBAB have a synergistic effect on the inhibition of zinc dendrites and have little effect on the discharge behavior of zinc electrodes. Banik et al. studied the effect of different contents of polyethylene glycol (PEG-200) on inhibiting the formation of zinc dendrites by in-situ observation [15]. When the concentration is 100 ~ 10,000 mg L^{-1}, the growth of zinc dendrite was inhibited by PEG-200. It was found that PEG-200 inhibited the growth of zinc dendrites by decreasing the exchange current density of the zinc electrodeposition reaction. Figure 5.1 shows the added concentration of 100 ~ 10,000 mg L^{-1}. The growth of zinc dendrite was inhibited by PEG-200.

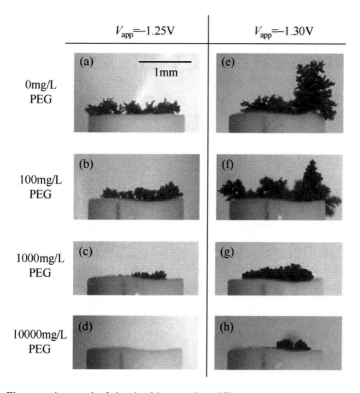

Fig. 5.1 Electron micrograph of zinc dendrite growth at different concentrations of PEG-200 [16]

5.1.2 Non-alkaline Liquid Electrolyte

5.1.2.1 Non-alkaline Electrolyte

Compared with alkaline Zn-air battery, neutral and slightly acidic Zn-air battery has the advantages of being cheap and easy to obtain electrolyte, low corrosiveness, and avoiding electrolyte carbonation. Although its working voltage and discharge current density are not as good as alkaline Zn-air batteries, it can meet the discharge requirements of medium and small current density and can replace alkaline Zn-air batteries in low-power discharge places.

About 5 mol L^{-1} NH_4Cl solution is used as the electrolytes for neutral and slightly acidic Zn-air batteries. Its discharge products are $Zn(NH_3)_2Cl_2$ with low solubility and it is easy to crystallize and plug the pores of the air electrode, thus reducing the battery working voltage and shortening the battery discharge time.

Z. H. Yu et al. studied the electrolyte solution of NH_4Cl, $ZnCl_2$, and KCl. It is found that the polarization of Cl solution in the zinc electrode in ammonium hydroxide was the lowest at low current density. At high current density, zinc electrode in $ZnCl_2$, the polarization is minimal in solution. Proper concentration of NH_4Cl solution as an electrolyte is beneficial to improving the activity of zinc electrodes. Adding KCl to the NH_4Cl solution can improve the conductive performance of the solution and improve the working voltage and discharge time of the Zn-air battery. The optimum electrolyte ratio of a quasi-neutral Zn-air battery is 4 mol L^{-1} NH_4Cl + 2 mol L^{-1} KCl.

5.1.2.2 Room Temperature Ionic Liquids

In the last decade, non-aqueous ionic liquids (ILs) have gradually become the ideal electrolyte for rechargeable Zn-air batteries. A nonaqueous ionic liquid, consisting entirely of ions (cations and anions), is a salt with a low melting point. Its melting point is around 100 °C or lower.

S. T. Deng et al. prepared [C14MIm][OH] by ion substitution reaction, and obtained ionic liquid crystal by mixing KOH solution to construct a high-performance solid-state liquid crystal electrolyte for Zn-air battery. In this study, bromotetradecane and 1-methylimidazole were used as raw materials to react at 65 °C to obtain [C14MIm][Br]. A methanol solution of KOH is prepared under dark conditions with a high concentration of OH^- Instead of Br^- The ion gets [C14MIm][OH]. KOH solution and [C14MIm][OH] were mixed by physical blending in a mass ratio of 50–70% to form Zn-air battery stream electrolyte. Under the polarized light microscope, the image shows that the mixture still retains the lamellar liquid crystal properties. Its layered LCD phase is presented between room temperature and 45 °C. At room temperature, the ion conduction capacity is between 5.89×10^{-3} S cm^{-1} and 2.59×10^{-2} S cm^{-1} with the quality hybrid ratio). In addition, [C14MIm][OH] changes significantly with temperature during cooling (Fig. 5.2).

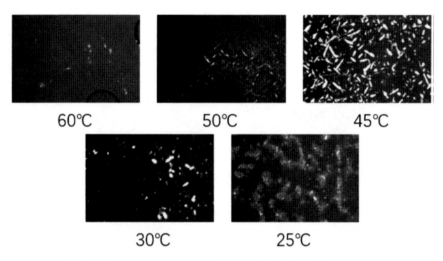

Fig. 5.2 Phase change of [C14MIm][OH] with temperature during cooling

Specifically, ionic liquids that melt at or below room temperature are defined as RTILs. Different selection and combinations of cations and anions lead to the diversity of RTILs. Therefore, the unique advantage of RTILs is their tunability, which makes them a good solvent and reaction medium in many research fields. The new RTILs applied to batteries has the following advantages. First, the inherent volume disadvantage of aqueous electrolytes is not a problem for non-aqueous RTILs, making them a safer alternative in batteries. At present, RTILs are widely used as an electrolyte in lithium-ion batteries to alleviate their safety problems. Second, complete ions are another feature that distinguishes RTILs from water electrolytes. Aqueous electrolytes originate from the dissolution of salts in aqueous solvents and are composed of dissolved ions, charged or neutral combinations, and solvent molecules. RTILs are salts that melt (liquefy) by providing heat to the system to overcome salt lattice energy, consisting only of ions and their combinations, without any molecular solvents. This ionic property guarantees a wide electrochemical window for the use of RTILs in batteries, as well as high electrode efficiency. The advantage of using aprotic RTILs in Zn-air cell systems is that zinc corrosion caused by hydrogen evolution can be avoided due to the absence of protons in the electrolyte. As a result, the current efficiency of electrodeposition of zinc in RTILs is improved, usually over 85%. Aprotic RTILs can also improve the shape of zinc deposition and prevent the dendritic formation of zinc. Studies have shown that RTIL cation influences deposition size and the morphology and growth direction of sediment are closely related to RTIL anion. In Zn-air battery systems, the development of RTIL-based electrolytes has achieved good results in reducing dendrite formation, avoiding electrolyte drying, and realizing reversible oxygen reactions. However, there are still some problems to be solved urgently. For example, the electrochemical dynamics of zinc and air electrodes in high-viscosity RTILs are nearly two orders of magnitude lower than in aqueous solution, which causes great difficulties in the practical application of RTILs.

5.2 Solid Electrolyte

In the past two or three decades, Zn-air batteries prepared by non-aqueous electrolytes have attracted extensive attention. Aqueous electrolytes can cause key problems such as electrode corrosion, dendrite formation, electrolyte drying, and leaching. Solid polymer electrolytes (SPEs) can solve the problem of stream electrolytes.

SPEs were introduced into Zn-air battery systems in the early 1990s to replace traditional alkaline aqueous electrolytes and separators. SPEs are ionic conductive solids formed by dissolving conductive salts into polymers. The main advantage of using SPEs in Zn-air batteries is the elimination of electrolyte leaching in fluid water systems, thereby increasing the battery life. In addition, the electrode corrosion problem in SPEs is also alleviated due to almost no convection.

5.2.1 Polymer Electrolyte

Polymer electrolytes such as polyvinyl alcohol (PVA) have a good hydrophilic ability, and superior film-forming ability, and are generally chosen as polymer bodies. Potassium hydroxide (KOH) has high electrical conductivity and is generally used as an ionic conductor of basic electrolytes. Gong et al. [17] assembled a flexible Zn-air battery for the first time in 2015 and prepared a hydrogel electrolyte with high ionic conductivity. They first dissolved polyoxyethylene (PEO) and PVA in water, then added KOH to improve electrical conductivity. After the film formation, the gel electrolyte has good mechanical properties, bending, and stretching capabilities. It can even stretch to 300% of the length without cracking. The resistance of the hydrogel electrolyte was 10 ω, and the ionic conductivity reached 0.3 S cm^{-1} by introducing KOH. Its water-storage capacity is not mentioned. Whether the performance changes after stretching, and whether it can return to the original shape is also a problem.

Other polymers such as polyacrylonitrile (PAN), polyacrylic acid (PAA), and polyacrylamide (PAM) are also commonly selected [18–21]. However, the combination of PVA-KOH or other polymers with KOH cannot meet the needs of building a high-performance flexible Zn-air battery. It is often necessary to introduce additional additives to improve the performance of the gel electrolyte. For example, Chen et al. [22] introduced SiO_2 into PVA. The particles not only enhance the ionic conductivity of the gel electrolyte but also improve the water retention and water absorption capacity of the electrolyte. Polyethylene glycol (PEG), PVA, and SiO_2 are heated, stirred, and cooled in the same solution to form a solid electrolyte film. SiO_2 with a mass fraction of 5%. As a plasticizer, the amorphous region of PVA increases, and the free molecular volume fraction increases, thus increasing the water absorption capacity of the electrolytes. At the same time, the test showed that SiO_2 was introduced. The ability of electrolytes to retain water was also improved. This is because the hydroxyl group on the surface of silica is conducive to the formation of hydrogen bonds, which further improves the water retention ability. On the other hand, the conductivity of the gel electrolyte increased from 36.1 mS^{-1} cm due to the addition of a 5% mass fraction of silica. It increased to 50.2 mS cm^{-1}. This is because the addition of silica filler contributes to the transformation of the PVA matrix from crystalline region to amorphous region, which ultimately leads to the improvement of ion migration ability.

Since the Zn-air battery is a semi-open system, the CO_2 in the air is easy to react with KOH in the gel electrolyte and forms potassium carbonate, making the ionic conductivity of the gel electrolyte sharply reduced. To improve the ability of electrolyte resistance to CO_2, Jiang et al. [23] combined poly (ethylene oxide) ($C_2H_6O_2$), polypropylene oxide ($C_6H_6O_2$), and poly (ethylene oxide) three-stage copolymerization into polymer electrolyte (F127). The electrolytes using F127 and 0.5 mol L^{-1} KOH and the commonly used 0.5 mol L^{-1} KOH electrolytes were assembled into Zn-air batteries, respectively, and were exposed to a CO_2 atmosphere for testing. The electrolytes of ordinary batteries react violently with carbon dioxide after only 24 h, while the open-circuit voltage of the batteries using F127 polymer does not change after 166 h, which significantly improves the ability to resist CO_2. This is because the viscosity of the F127 polymer is 107 times higher than that of 0.5 mol L^{-1} KOH at room temperature. Higher viscosity means a lower diffusion coefficient of CO_2.

In addition to enhancing the anti-CO_2 performance, the gel electrolyte also needs to improve the ionic conductivity. Sun et al. [24] used tetraethylammonium hydroxide to replace the traditional potassium hydroxide as the ionic conductor, denoted as tetraethylammonium hydroxide (TEAOH). Compared with the commonly used KOH-PVA, TEAOHPVA gel electrolyte has significantly improved water retention capacity and ionic conductivity.

To expand the range of flexible Zn-air batteries, gel-type electrolytes should not only face strong alkali solutions but also maintain good electrochemical and mechanical properties at high and low temperatures. Therefore, Glycerol was introduced into the hydrogel by Jiang et al. [23]. It formed hydrogen bonds with water while reducing its saturated vapor pressure, resulting in a gel electrolyte that operates from − 20 to 70 °C. They combined PAM with PAA and methylene bisacrylamide (Gly), while Gly and ammonium persulfate (APS) were also added. The structure presents a cross-linked network. The obtained electrolyte has good ductility. Normal PAM + PAA does not show ductility. With the addition of Gly, the mechanical strength first increases to the maximum value and then decreases. The maximum stress is 145.5 kPa and the breaking tension is 484.4 kPa. Adding too much glycerin will degrade the mechanical properties of the gel electrolyte. A Gly-free hydrogel becomes a hard white plate at − 20 °C. The hydrogel electrolyte with Gly was transparent, soft, and stretchable, without an obvious freezing phenomenon. Therefore, it can be concluded that Gly makes hydrogel possess high flexibility and ductility below 0 °C. The charge–discharge polarization curves of the assembled flexible Zn-air battery do not change significantly at different bending angles. The voltage variation is also smaller after the bending and hammering tests. Up to now, people have been able to improve the performance of polymer gel electrolytes by introducing various additives, but the perfect polymer gel electrolyte has not been developed. Good polymer gel electrolyte should meet the following characteristics: high flexibility, high water absorption capacity, and water retention capacity, high ionic conductivity, wide operating temperature range, CO_2 resistance ability, and good fitting performance of anode and cathode of the battery.

5.2.2   Cellulose Gel Electrolytes

Polymer electrolyte has poor stability due to its high hydrophilicity, and it is easy to deform. At the same time, alkali will be released periodically, leading to a sharp decline in performance. To solve the above problems, it has been found that nanocellulose contains many hydroxyl functional groups. It has the characteristics of good flexibility, high volume ratio, and large specific surface area, which can make up for the shortcomings of polymer electrolytes.

An innovative gel electrolyte combining sodium polyacrylate with cellulose was developed [25]. By cross-linking cellulose, polyacrylonitrile, and methylene bisacrylamide (MBAA), they created porous networks, dissolved in oxygen and strong in toughness. With the addition of PANa, the conductivity of the gel electrolyte increased from 0.15 to 0.28 S cm^{-1}. At the same time adding 2.7% cellulose, the breaking strength and tensile strength of the electrolyte were greatly improved, and the tensile strength was 1000%. Using only PANa electrolyte, the tensile strength is only 300%. In alkaline solutions such as PAA, PAM, and PANa, the hydrogel network is extremely easily broken due to OH^- easy interacting with $-COOH$ or $-NH_2$ in a hydrogel. The usual solution is to add NaOH as a neutralizer. This inevitably increases the volume and reduces the tensile strength of the gel electrolyte. However, the cellulose chain formed by adding cellulose can effectively inhibit the swelling of hydrogel and significantly improve the mechanical properties. Therefore, the sandwich made by L. Ma et al. and the fiber Zn-air battery can reach the tensile strength of 800% and 500% respectively. After stretching, the charge–discharge cycle performance of the battery did not change significantly.

In addition, the covalently crosslinked polymer network of the covalent system can ensure sufficient hardness and toughness of the hydrogel [26]. Meanwhile, the non-covalent crosslinked bond dissipates energy during loading, which improves the fatigue resistance and recovery ability of the double-network hydrogel. W. Sun et al. developed an alkaline double network cellulose gel electrolyte. They combined poly (2-acrylami-2-methylpropanesulfonic acid potassium salt) (PAMPS-K) with interpenetrating methylcellulose (MC). The hydrogel retains considerable mechanical strength and ionic conductivity even at − 20 °C. At 25 °C, 5 mol L^{-1} KOH, the ionic conductivity reaches 105 mS cm^{-1}. The specific capacity density of the gel Zn-air battery was 764.7 mAh g^{-1} and the energy density reaches 850.2 mWh g^{-1}.

5.2.3 Graphene Oxide (GO) Gel Electrolytes

GO has been demonstrated to have high ionic conductivity and chemical stability, as well as several electron insulators containing oxygen groups (epoxy, hydroxyl, and carboxyl), which can be easily functionalized to improve its ionic conductivity [27]. Therefore, cellulose and GO were first functionalized with quaternary ammonium (QA) by H. Zhang et al. and then stacked in layers, and ion-exchanged. The treated GO and cellulose contained large amounts of OH^{-} ions, which further improved the conductivity of the electrolyte (58.8 mS cm^{-1}, 70 °C). Ordinary GO is only 2.5 mS cm^{-1}. The cross-sections of QAFCGO films were very uniform and smooth. The tightly interwoven network structure of cellulose also shows good tensile properties of the electrolyte. When assembled into a Zn-air battery, its stability and circulation capacity are superior to that of the commercial A201. Its open circuit voltage is 1.41 V, similar to the commercial A201. Current density exceeds 20 mA cm^{-2}, and QAFCGO battery charge and discharge overcurrent potential are significantly smaller than commercial A201. In the 60 mA cm^{-2}, The charging potential of QAFCGO is 291 mV smaller than that of A201, and the discharge potential is 154 mV smaller than that of commercial A201. The current density is 1 mA cm^{-2}, QAFCGO cycle 30 times (10 min discharge and 10 min charge into a cycle), while commercial A201 cycle only 15 times.

Based on the above work, J. Song et al. initiated a new direction by combining GO with polymers. They mixed PVA, PAA, GO, and KI. After modification by PVA, they constructed a crosslinked composite network structure. Due to the abundance of oxygen functional groups on the GO surface and the PVA chain, PVA and GO can be cross-linked by forming hydrogen bonds. Its operating time reaches 200 h. Compared with a PVA-based Zn-air battery, KI-PVA based Zn-air battery can discharge for 166 h. Similarly, flexible Zn-air batteries assembled using KI-PVA are flexible and textile and can be used to charge various electronic products [28]. To sum up, the current mainstream electrolyte preparation method is to combine polymer, cellulose, GO, with other substances to prepare high-performance electrolytes. However, electrolytes that meet the requirements of mechanical properties, high and low-temperature resistance, high ionic conductivity, and so on have not been born yet, and are in urgent need of exploration.

5.2.4 High Concentration Zinc Ion Electrolyte

In 2015, researchers found that high salt concentration (water in salt) could greatly improve the electrochemical performance of stream electrolytes. "Water in Salt" electrolyte is an aqueous solution with 21 mol kg^{-1} lithium trifluoromethane sulfon-imide (LiTFSI). The density of LiTFSI is much more than water. M. Wang et al. developed a high-concentration zinc ion electrolyte (HCZE, 1 mol L^{-1} Zn(TFSI)$_2$ + 20 mol L^{-1} LiTFSI) for zinc-based batteries based on the concept of "water in salt". The electrolyte prevents water loss and is highly reversible, enabling dendrite-free plating/dissolution of zinc with a Coulomb efficiency of nearly 100%. The high concentration of anion TFSI- and Zn^{2+} in the electrolyte form a tight solvent-sheath structure ion pair (Zn-TFSI)$^+$, which inhibits the generation of (Zn-(H$_2$O)$_6$)$^{2+}$. The Zn-air battery with HCZE as the electrolyte can be cycled more than 200 times, and the battery power can reach 300 Wh kg^{-1} [29]. This efficient zinc utilization method can be extended to other battery systems with slow reaction kinetics.

L. Chen et al. continued to reduce the water content in the electrolyte based on "water in salt", and developed a room temperature zinc molten hydrate electrolyte, ZnCl$_2$·nH$_2$O. All the water molecules are associated with Zn^{2+} to form a hydration layer. When it is used as the electrolyte of a rechargeable Zn-air battery, there is no side reaction caused by excess water, which eliminates the disadvantages of water-electrolyte. A high concentration of zinc can slow down the formation of a concentration gradient, and a stable ion supply can ensure regular deposition to prevent dendrite formation. Experiments show that ZnCl$_2$·2.33 H$_2$O has the lowest activity and the most hydration. At room temperature of 30 °C, the Zn-air battery assembled with Pt/C electrode as the positive electrode, Zn as the negative electrode, and ZnCl$_2$·2.33 H$_2$O as the electrolyte can achieve a reversible capacity of 1000 mAh g^{-1} in 100 cycles. The coulombic efficiency of charge and discharge reaches 99% and no zinc dendrite is produced. Unlike organic fluoride which is expensive, this electrolyte material is easy to obtain and low manufacturing cost, making it an ideal substitute for existing electrolytes.

References

1. E.D. Farmer, A.H. Webb, J. Appl. Electrochem. **2**, 123 (1972)
2. N.A. Hampson, M.J. Tarbox, J. Electrochem. Soc. **110**, 95 (1963)
3. T.P. Dirkse, N.A. Hampson, Electrochim. Acta **16**, 2049 (1971)
4. T.P. Dirkse, N.A. Hampson, Electrochim. Acta **17**, 387 (1972)
5. M.B. Liu, G.M. Cook, N.P. Yao, J. Electrochem. Soc. **128**, 1663 (1981)
6. S. Thomas, I.S. Cole, M. Sridhar, N. Birbilis, Electrochim. Acta **97**, 192 (2013)
7. R.W. Powers, M.W. Breiter, Electrochem Soc-J **116**, 719 (1969)
8. H. Yang, Y. Cao, X. Ai, L. Xiao, J. Power Sources **128**, 97 (2004)
9. A.R. Suresh Kannan, S. Muralidharan, K.B. Sarangapani, V. Balaramachandran, V. Kapali, J. Power Sources **57**, 93 (1995)
10. J. Dobryszycki, S. Biallozor, Corros. Sci. **43**, 1309 (2001)
11. M. Naja, N. Penazzi, G. Farnia, G. Sandonà, Electrochim. Acta **38**, 1453 (1993)
12. J.I.A. Zhe, Trans. Nonferrous Metals Soc. China **15**, 200 (2005)
13. S.J. Banik, R. Akolkar, J. Electrochem. Soc. **160**, D519 (2013)
14. K. Zhang, R. Simic, W. Yan, N.D. Spencer, ACS Appl. Mater. Interfaces **11**, 25427 (2019)
15. P. Karami, C.S. Wyss, A. Khoushabi, A. Schmocker, M. Broome, C. Moser, P.E. Bourban, D.P. Pioletti, A.C.S. Appl, Mater. Interfaces **10**, 38692 (2018)

16. C.Y. Chen, K. Matsumoto, K. Kubota, R. Hagiwara, Q. Xu, Adv. Energy Mater. **9**, 1900196 (2019)
17. J.P. Gong, Y. Katsuyama, T. Kurokawa, Y. Osada, Adv. Mater. **15**, 1155 (2003)
18. G. Fotouhi, C. Ogier, J.-H. Kim, S. Kim, G. Cao, A.Q. Shen, J. Kramlich, J.-H. Chung, J. Micromech. Microeng. **26**, 055011 (2016)
19. X. Fan, J. Liu, Z. Song, X. Han, Y. Deng, C. Zhong, W. Hu, Nano Energy **56**, 454 (2019)
20. S. Zhao, D. Xia, M. Li, D. Cheng, K. Wang, Y.S. Meng, Z. Chen, J. Bae, A.C.S. Appl, Mater. Interfaces **13**, 12033 (2021)
21. M. Li, B. Liu, X. Fan, X. Liu, J. Liu, J. Ding, X. Han, Y. Deng, W. Hu, C. Zhong, ACS Appl. Mater. Interfaces **11**, 28909 (2019)
22. R. Chen, X. Xu, S. Peng, J. Chen, D. Yu, C. Xiao, Y. Li, Y. Chen, X. Hu, M. Liu, H. Yang, I. Wyman, X. Wu, ACS Sustain. Chem. Eng. **8**, 11501 (2020)
23. G. Jiang, M. Goledzinowski, F.J.E. Comeau, H. Zarrin, G. Lui, J. Lenos, A. Veileux, G. Liu, J. Zhang, S. Hemmati, J. Qiao, Z. Chen, Adv. Funct. Mater. **26**, 1729 (2016)
24. N. Sun, F. Lu, Y. Yu, L. Su, X. Gao, L. Zheng, A.C.S. Appl, Mater. Interfaces **12**, 11778 (2020)
25. W. Gao, G. Wu, M.T. Janicke, D.A. Cullen, R. Mukundan, J.K. Baldwin, E.L. Brosha, C. Galande, P.M. Ajayan, K.L. More, A.M. Dattelbaum, P. Zelenay, Angew. Chem. Int. Ed. **53**, 3588 (2014)
26. J. Zhang, J. Fu, X. Song, G. Jiang, H. Zarrin, P. Xu, K. Li, A. Yu, Z. Chen, Adv. Energy Mater. **6**, 1600476 (2016)
27. J. Lim, K. Park, H. Lee, J. Kim, K. Kwak, M. Cho, J. Am. Chem. Soc. **140**, 15661 (2018)
28. F. Wang, O. Borodin, T. Gao, X. Fan, W. Sun, F. Han, A. Faraone, J.A. Dura, K. Xu, C. Wang, Nat. Mater. **17**, 543 (2018)
29. H. Liu, Y. Liu, J. Li, Phys. Chem. Chem. Phys. **12**, 1685 (2010)

Chapter 6
Applications of Zinc-Air Batteries

6.1 Application Background

Since the outbreak of the industrial revolution, capitalism has completed the transformation from farm handicrafts to machine industry, and the consequent large-scale use of large industrial machines has led to a dramatic increase in global demand for energy. The world's economy and technology have developed to an unprecedented extent. At the same time, the burning of large amounts of fossil fuels also pollutes the air and causes irreparable damage to the environment. In the London smog incident of 1952, for example, on 5 December, an anticyclone over London caused a large number of emissions from factories and coal heating to accumulate over the city, making it difficult to spread. London was enveloped in thick smog, bringing traffic to a standstill as pedestrians groped their way cautiously. Not only were people's lives disrupted, but their health was also seriously compromised. Many people suffered from chest congestion, suffocation, and other discomforts, and morbidity and mortality rates increased sharply. According to statistics, as many as 4000 people died that month because of dense fog. It was not until 9 December that a strong, cold westerly wind blew away the smog that had enveloped London and the disaster was lifted. The environmental problems caused by using fossil fuels, such as acid rain, melting of the polar glaciers due to the greenhouse effect, destruction of the ozone layer, and atmospheric pollution, are also major environmental problems that the world has had to solve since industrialization. For this reason, since the last century, countries have been exploring the use of clean and harmless natural energy sources such as solar, wind, tidal and hydro energy. According to scientists, in 2050, the global outlook for power generation will be 62% renewable energy [1]. In addition, our country is also vigorously promoting new energy vehicles, using new energy instead of traditional energy sources, believing that the development of new green energy vehicles is a necessary path for China to become a strong automotive nation, and is an important step to better improve environmental pollution and vigorously develop the economy. However, the use of new energy sources is still facing many problems, including low energy conversion efficiency and difficulties in storage and transportation. For this

S. Peng, *Zinc-Air Batteries*, https://doi.org/10.1007/978-981-19-8214-9_6

reason, many scientists have been exploring efficient energy storage and conversion devices to solve these problems, including electrolytic water, metal-air batteries, fuel cells and so on.

The metal-air battery, a type of chemical battery, is not identical in composition to a dry cell, the difference being that the oxidizer in an air battery comes from oxygen in the air and the negative electrode generates electricity through a chemical reaction with the oxygen in the air. As the earth is now covered with oxygen, the source of oxygen for an air battery is theoretically unlimited and the storage capacity of an air battery can be 5–10 times that of an ordinary battery. The current metal-air battery has been initially used in the field of electric cars and portable electronic devices and has the advantages of being environmentally friendly and relatively safe.

The raw materials available for making air batteries are relatively diverse, and the metal-air batteries that have made progress in research are mainly aluminium-air batteries, magnesium-air batteries, zinc-air batteries, and lithium-air batteries, etc. Lithium-air batteries have higher specific energy and voltage, and can provide high power and sufficient power. However, lithium batteries are expensive to develop, and safety issues still need to be improved. For example, lithium in metallic form is plagued by its inherent instability when exposed to air and liquid electrolytes, which may affect the performance of the battery. In recent years, accidents regarding lithium batteries have occurred, and in addition to the problems caused by the unregulated operation of users, there is no shortage of miscellaneous lithium plants that cut corners to save costs and do not have even the most basic protection devices, and these situations This makes the use and promotion of lithium batteries difficult. Magnesium-air and aluminium-air batteries are both compatible with liquid electrolytes and have energy densities comparable to lithium-air batteries; however, due to their lower reduction potential, they can lead to rapid self-discharge and lower coulombic charging efficiency through hydrogen precipitation reactions. Zinc-air batteries have been developed earlier, have a wider range of applications, have improved safety compared to lithium-ion batteries, and are currently a hotter metal-air battery in terms of both application and research. This chapter will take zinc-air batteries as an example and briefly introduce the background of their application as well as knowledge of where they are used.

Zinc-air batteries, also known as zinc-oxygen air batteries, have a history of more than 100 years of development, a small, high charge capacity, small mass, can work in a wide range of temperatures, non-corrosive and safe and reliable work of environmentally friendly batteries.

The electrochemical reactions occurring at the anode and cathode during the discharge of a zinc-air cell are:

$$Zn + 4OH^- \rightarrow Zn(OH)_4^{2-} + 2e^-$$
$$Zn(OH)_4^{2-} \rightarrow ZnO + 2OH^- + H_2O$$
$$O_2 + 2H_2O + 4e^- \rightarrow 4OH^-$$

The total electrochemical reaction is

$$2Zn + O_2 \rightarrow 2ZnO$$

There are many advantages of zinc-air batteries, for example: (1) the specific energy is large, because the active material used in the air electrode is oxygen in the air, i.e. the active material is outside the battery, so the theoretical specific energy of the air battery is much larger than that of the general metal oxide electrode. Its specific energy is 4–6 times greater than that of lead-acid batteries; (2) simple manufacturing process, low cost; (3) safe and reliable, even if the external encounter open fire, short circuit, puncture, impact, etc. will not occur in the event of combustion, explosion; (4) environmental protection zinc-air batteries used in the positive electrode is a carbon rod, the negative electrode is used in the metal zinc, positive and negative electrodes are not used toxic substances. (5) The raw materials are easily and cheaply available, and there are no special difficulties or dangers in their use.

Zinc-air batteries can be classified according to different criteria as follows.

1. Nature of the electrolyte: they can be divided into slightly acidic and alkaline batteries.
2. The way the electrolyte is handled: static and cyclic batteries.
3. Shape of the battery: can be divided into rectangular, button-type, and cylindrical batteries.
4. Form of supply of air: can be divided into internal oxygen type and external oxygen type batteries. Internal oxygen type refers to the negative plate of the battery around the two sides of the positive gas electrode, the battery has a complete shell; external oxygen type refers to the negative plate of the battery in the middle of the positive gas electrode, the gas electrode doubles as part of the outer wall of the battery shell [2].
5. According to the negative electrode charging form to divide: can be divided into mechanical charging type batteries, primary batteries and secondary batteries. Zinc-air batteries can be mechanically charged by removing the used-up zinc from the battery and replacing it with a new electrode material. This avoids the problems caused by the poor reversibility of the zinc electrode and the instability of the air electrode and ensures that the zinc air battery continues to work. Waste zinc can also be recycled, which helps to protect the environment and save resources. However, the short service life, high costs, and intermittent operation have stopped the development of rechargeable zinc air batteries. Primary batteries have a long and stable storage life and a high discharge capacity. Sealed batteries will only show low-capacity degradation after one year of storage; like other primary batteries, they can be combined in series to produce devices with higher voltages for high power. Secondary zinc-air batteries can be recharged, allowing for much higher utilization of zinc-air batteries.

Due to these advantages, experts consider zinc-air batteries to be the most suitable power source for use in equippable power equipment. Zinc-air batteries have a voltage of around 1.4 V. The discharge current is governed by the oxygen adsorption and diffusion rate of the activated carbon electrode. Each battery type has its optimum current value, but if this limit is exceeded then the activated carbon electrode deteriorates rapidly, resulting in a reduction in battery performance. The charge of a battery is generally more than three times greater than that of a zinc-manganese battery of the same volume. Large zinc-air batteries have a charge of 500–2000 Ah and are mainly used in railway and maritime beacon devices. Button-shaped zinc-air batteries have a charge of 200–400 mAh and have been widely used in hearing aids. If the use of zinc-air batteries could be widespread, then air and noise pollution could be greatly reduced. Zinc reserves in the earth's crust are three times larger than lithium, and China's zinc reserves are the largest in the world, which can provide a large amount of raw material support for zinc-air batteries. At a time when the difficulties faced in the commercialization of hydrogen–oxygen fuel cells, such as high prices, huge investments and technical bottlenecks, are momentarily insurmountable, the upgraded use of zinc-air batteries provides a shortcut to the popularization of air batteries and is expected to play a major role in the storage of new energy sources. The use of zinc-air batteries is expected to play an important role in the storage and utilization of new energy sources, solving current energy and environmental problems.

6.2 Application Sites

Zinc-air batteries are diverse and perform well in many ways, so zinc-air batteries are used in a wider range of scenarios than delivery. For example, zinc-air batteries can be used in automobiles to replace traditional cars. In portable devices, such as in hearing aids, mobile phones, bracelets, some wearable electronic devices, some fixed devices. And such as zinc-air batteries as an energy storage candidate material, can be used to solve the instability of wind and solar power and the periodic fluctuations of the power grid, helping to improve the energy efficiency of the power grid. It can also be used to optimize load distribution on the grid. Through the rational modification and optimization of different battery components, researchers have gradually improved the efficiency of battery use and extended its service life, both reducing assembly costs and bringing the application of zinc-air batteries closer and closer to our lives.

6.2.1 Applications in the Portable Sector

As early as the beginning of the twenty-first century, great achievements have been made in the practical research of alkaline zinc-air batteries abroad. They can be used in large-scale automobile fields, and can also enter portable electronic products, such as hearing aids, electronic watches, calculators, electronic dictionaries and so on. For many people in the modern world, compact and energy-efficient items are increasingly popular. Represented by the Israeli Electric Fuel Company and the American Evonyx Company [3], it is the first to use zinc-air batteries in the field of mobile communications. They have achieved great success in alkaline traction batteries, which means that environmentally friendly zinc-air batteries will gradually replace the environmentally harmful Lead storage battery as a development trend.

A hearing aid is an electronic device used to amplify sound and compensate for hearing loss. It usually consists of a microphone, amplifier, headphones and is powered by a low-voltage battery. The converter consists of a microphone (microphone or microphone), a magnetic induction coil and other components, whose main function is to convert the received acoustic signal into an electrical signal for transmission to the signal conditioning unit. Simple analog amplification circuits are used, which only amplify the electrical signal in a fixed proportion; the signal conditioning unit amplifies the electrical signal by a factor of 10,000 or even tens of thousands and passes it on to the output converter [4, 5]. The output converter is usually a headset, which converts the amplified signal from electrical energy to acoustic energy, and the power supply is primarily aimed at driving the hearing aid. To facilitate and better serve deaf people, it is vital to choose the service life and model of the hearing aid, and the battery directly affects the normal working time of the hearing aid, as well as the gain and degree of hearing loss. The longer the battery is used during normal use, the longer the interval between battery replacement by the patient and the less chance of accidents. Hearing aid battery use instructions and precautions hearing aid disposable batteries, generally must have complete packaging to ensure that the batteries cannot contact the air before use, to avoid the occurrence of battery power loss. The batteries should be used at low power levels as far as possible to ensure a certain working life and avoid frequent replacement. When the hearing aid battery is exhausted or disabled, remove the battery immediately to avoid battery leakage damage to the hearing aid or accidental injury to the ear. After removing the battery, place the battery in a sunny or dry environment to prevent the battery from exploding. While small button zinc-air batteries were designed from the very beginning to power hearing aids, their structure is shown in Fig. 6.1a. Figure 6.1b–d [4] shows a structure of disposable button zinc-air battery currently used commercially for small hearing aids. The development of zinc-air batteries has played an important role in the widespread use of hearing aids. Batteries of the "button" construction are marketed in large numbers and market research shows that zinc-air batteries and silver oxide, mercury oxide and manganese dioxide button batteries are the most popular.

As far as mobile phone batteries are concerned, the performance of zinc-air batteries is more significant than previous nickel-metal hydride batteries and lithium-ion batteries. Under normal circumstances, the talk time is more than 5 times higher than that of other batteries, and the standby time can reach more than 50 days. The first trial of zinc-air batteries in mobile phones was carried out by the Israeli company electro-fuel, mainly for Motorola mobile phones, where continuous talk time of 6.2 h was achieved during the trial phase. A square zinc-air battery pack for mobile phones was also developed by Wuhan University in China, as shown in Fig. 6.2 [4]. This battery pack has a specific energy of 220 Wh kg^{-1} and a maximum output power of 3.6 W, which can meet the comprehensive performance standards for mobile phones [4].

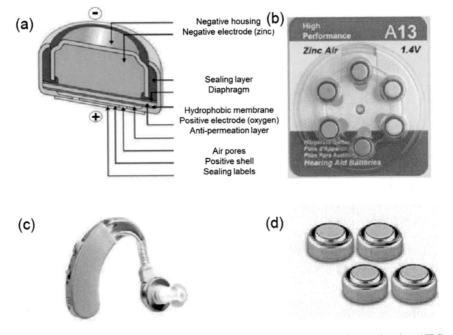

Fig. 6.1 a Zinc-air batteries structure for hearing aids; **b–d** Zinc empty battery hearing AIDS commonly available on the market

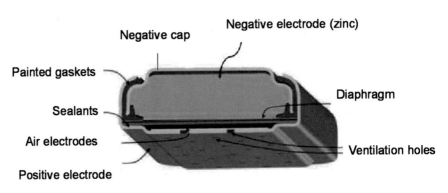

Fig. 6.2 Zinc-air battery for mobile phones

6.2.2   Application in the Field of Electric and Hybrid Vehicles

With the rapid development of society, people's living standards and purchasing power are gradually improved, and people are no longer satisfied with simple subsistence, but gradually shift consumption to how to eat more nutritious, travel more convenient and fast, so the car gradually becomes people's means of transportation.

But in the face of the increasingly serious energy and environmental problems, automotive technology towards fuel diversification, power electrification of new energy direction change. In its long history, mankind has experienced two reforms of energy and power systems, each of which has brought great changes to the world and enabled the technology and economy of the pioneer countries to take off. Now, mankind is once again at the crossroads of energy power reform—replacing the internal combustion engine with electricity and power batteries. Because of their energy-saving and environmentally friendly features, new energy electric vehicles have become the object of vigorous development in various countries. The development of new energy-electric vehicles is widely regarded as an important way to reduce energy losses and transform economies [6]. As the "engine" component of electric vehicles, the power battery plays a key role, and many scholars and institutions, both domestic and foreign, are now actively working on the research and development of power batteries [7].

Whether it is the pressure of environmental protection and energy depletion or the active pursuit of drive system changes for development, new energy-powered vehicles will undoubtedly become the future direction of automotive development. If new energy vehicles are developed rapidly, China will have 140 million vehicles in 2020, saving 32.29 million tons of oil and replacing 31.1 million tons of oil, saving, and replacing a total of 63.39 million tons of oil, equivalent to a 22.7% reduction in the demand for oil for automobiles. By 2030, the development of new energy vehicles will save 73.06 million tons of oil and replace 91 million tons of oil, saving and replacing a total of 164.06 million tons of oil, equivalent to a 41% cut in automotive oil demand [8]. At that time, biofuels and fuel cells will play an important role in automotive oil replacement.

So far, lithium-ion batteries are commonly used in electric vehicle power batteries. Lithium-ion batteries have the advantages of high specific energy, high specific power, and a wide operating range. But at the same time there are also shortcomings, such as the charging current cannot be too high, the cycle life is short, and the working and transportation process is flammable and explosive. To find a safer and more environmentally friendly battery for electric vehicles, metal air batteries have started to enter people's vision. Metal-air batteries are a new type of battery between traditional batteries and fuel cells. Metal-air batteries have many advantages. Their positive active material is oxygen, and the supply of oxygen in the air is very sufficient. Theoretically speaking, if there is a constant supply of oxygen entering, they can discharge indefinitely.

Among the various types of metal-air batteries, zinc-air batteries have a relatively high energy density, with a theoretical mass specific energy of up to $1353\,\mathrm{Wh\,kg^{-1}}$ [4]. Zinc raw material has abundant reserves, low cost, environmental protection, transportation and use in the process of no explosion and other significant advantages. The zinc-air batteries currently used in electric vehicles mainly include rechargeable zinc-air batteries and mechanically rechargeable zinc-air batteries (zinc-air fuel cells). Rechargeable zinc-air batteries are subject to problems such as dendrite growth, zinc electrode deformation and passivation during the charging process, leading to a decrease in the performance of the negative electrode, which to a certain extent hinders the development of rechargeable zinc-air batteries. Given the difficulties in solving the problems of rechargeable zinc-air batteries, a new "mechanical charging" concept has been developed, whereby the zinc electrode is mechanically replaced to complete the charging process. Mechanically rechargeable zinc-air batteries do not have the problem of zinc dendrite growth during the charging process [9].

In 1995, the Israeli company Electric Fuel used zinc-air batteries for the first time in electric vehicles, using mechanical replacement of new motors to charge the current, with a specific energy of up to 175 Wh kg^{-1}, and successfully applied to the MB410 postal car (produced by Mercedes-Benz) of the German postal system, with a maximum speed of 120 km h^{-1}. The replacement of zinc electrodes and zinc electrode regeneration work has a special charging station to complete. DEMI in the USA and several other countries such as Germany, France, Sweden, the Netherlands, Finland, Spain, and South Africa are also actively promoting the use of zinc-air batteries in electric vehicles.

6.2.3 The Field of Wearable Flexible Batteries

In recent years, with the development of flexible wearable electronics, the demand for flexible energy storage devices with high energy density and light weight has been increasing. Researchers have developed a variety of flexible energy storage devices based on new nanomaterials, such as flexible supercapacitors and lithium-ion batteries with novel features such as planar stretchability and wire weave ability. However, the light and thin cell structure design for flexible requirements limits the amount of electrode active materials used, resulting in existing flexible energy storage devices with low energy densities that are not well suited to practical use. Zinc-air batteries have a high theoretical energy density, safe and environmentally friendly, and more suitable for close use scenarios. Therefore, the development of flexible rechargeable zinc-air batteries has become a research hotspot, and many milestones have been achieved recently.

Flexible electronic devices require that they can maintain good working performance under conditions of high mechanical stress generating large deformation. In recent years, the emergence of foldable displays, electronic skin, implantable medical devices, etc., representing the future direction of development of intimate and implantable electronic products. Traditional energy storage devices have brittle electrodes that are prone to bending and splitting, and the use of liquid electrolytes leads to bulky packaging shells that cannot meet the flexibility requirements of flexible electronic devices. Therefore, the development of light, thin, efficient, safe, and economical flexible energy storage devices is imminent, and flexible rechargeable zinc-air batteries are a promising alternative drive power source for wearable electronic devices. Unlike conventional forms of energy storage such as supercapacitors and lithium-ion batteries, the semi-open structure of zinc-air batteries makes their flexible design more constrained and therefore faces greater challenges in material preparation to achieve flexible air cathodes.

At present, flexible zinc electrode materials mainly include zinc monomer (such as zinc wire and zinc foil) and zinc loaded on a flexible substrate. Flexible air electrodes can achieve electrode thinning through functional integration, and flexible solid/semi-solid electrolytes are mainly alkaline gel electrolytes. There are two main structures of flexible zinc-air batteries that have been reported: planar and linear [10].

1. Planar type flexible zinc-air batteries

   The conventional zinc-air battery components are thinly sliced and flexible, and then stacked in sequential layers to form a planar-type flexible zinc-air battery, which is the most common solid-state flexible zinc-air battery configuration.

   J. Fu et al. in the University of Waterloo laminated a PVA gel electrolyte between an air electrode and a zinc electrode to obtain a planar flexible zinc-air cell [11]. The porous polymer electrolyte film was obtained by the reverse transfer method, and the overall thickness of the flexible Zn-Air cell was 0.5 mm with a carbon cloth loaded with a bifunctional catalyst as the air electrode, a zinc sheet as the zinc electrode and a copper foil as the substrate collector.

2. Linear flexible zinc-air battery

   It is suggested that the linear flexible zinc-air battery is a flexible battery with a linear or fibrous shape. Commonly, the linear flexible zinc-air battery has a spring-shaped or wire-shaped zinc wire as its core, which is covered with a gel electrolyte, and the electrolyte layer is then wrapped or wound with a flexible air electrode. Wire-type structures are typically more flexible than planar zinc-air batteries and can be braided and knotted for wearable electronic device applications, as shown in Fig. 6.3 [12].

Joohyuk Park et al. [13] designed a wire-type flexible zinc-air battery. This battery has a spring-like zinc foil as the center of the circle, a gelatin electrolyte poured, a carbon cloth loaded with a non-precious metal bifunctional catalyst wound as the air electrode, and finally a porous heat shrink tube wrapped as the gas diffusion channel. This linear flexible zinc-air battery discharges at 0.92 V for 9 h and can maintain the same discharge voltage in the bent state as in the non-bent state when discharged at a constant current of 0.1 mA cm^{-2} at room temperature in atmospheric conditions [10].

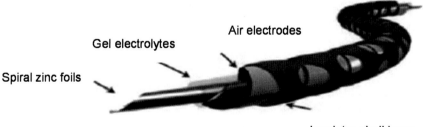

Fig. 6.3 Schematic diagram of a wire-type flexible rechargeable zinc-air battery

In addition, there are some extended developments of flexible batteries, such as transparent flexible zinc-air batteries, weaveable flexible zinc-air batteries, stretchable flexible zinc-air batteries, etc. In summary, significant progress has been made in the development of flexible zinc-air batteries in the laboratory in recent years. Several different structures of flexible zinc-air batteries have been developed to improve mechanical stability under different deformations while maintaining good electrochemical performance, including high discharge capacity, high power density, high energy efficiency and long cycle life. However, many technical challenges remain and breakthroughs in several key areas are needed to develop a new generation of flexible Zn-Air batteries with significantly improved battery performance, especially durability [14].

6.2.4 Applications in the Field of Energy Storage

As ecological problems such as energy shortages and environmental degradation become increasingly serious, efforts to find and develop new, clean, and sustainable sources of energy have become a crucial task at this stage. To achieve a new era of sustainable development that is green, healthy, economical, and environmentally friendly, some renewable energy sources, such as solar, wind, tidal and hydropower, are promising new alternatives to the use of traditional fossil fuels for power generation. They are clean and environmentally friendly, simple, and easy to obtain, and can achieve zero pollution to the environment [15].

However, the current situation is that the output of electricity produced using this new clean energy varies greatly in terms of season, climate, and geography, and cannot supply electricity in a particularly stable manner, often failing to meet the huge energy demand of people in today's society, and this unstable output of electricity is also highly likely to affect and shorten the life of the grid. In China, for example, due to geographical factors, many wind farms are in the western desert, but there are no nearby lines that can be used to send electricity out, and the practice of setting up special lines for wind power frames is not scientific. This is because even if the wind is converted into electricity, the losses make it less than the normal fossil fuel supply. In addition, due to the high cost of unstable wind energy in terms of loss of grid life and later maintenance and repair, to make the most of it, grid companies are now using battery banks to store energy in newly built wind farms. But battery capacity makes it difficult to provide large amounts of power on a sustainable basis. Beijing has ordered wind power operators to stop expanding wind capacity up to four times over the past five years [16], because unreliable wind power is damaging the national grid and costing the government large sums of money. The best areas to deploy wind turbines in China are far from the most populated coastal provinces, and building the infrastructure to transmit wind power over long distances is very expensive, amounting to several times the cost of generating electricity. And because wind turbine generation is intermittent and does not consistently match the times of day when electricity is most needed, it poses a huge challenge to grid operators and makes the grid very vulnerable. Therefore, researchers are working to develop new materials that are green and sustainable while at the same time conforming to the concept of providing a stable supply of electrical energy.

The advantages of lithium-ion batteries as an environmentally friendly battery with high electrochemical capacity, renewable, higher energy density, and good cycling performance make lithium-ion batteries the main choice today, occupying a large part of the market for electric vehicles and will remain in the most demanded position for some time to come, and will undergo a series of improvements and developments. However, the development of lithium-ion batteries is still a safety concern and the high price of lithium resources due to the low lithium content of the earth is also a reason for limiting the development of lithium-ion batteries.

To sum up, the use of lithium batteries as energy storage equipment for power plants is not only a certain risk, and the cost of construction is relatively high. In recent years some air batteries (mainly lithium-air batteries and zinc-air batteries) because of its unit energy density (including weight energy density and volume energy density), and the principle of metal air batteries gradually came into the people's vision. Zinc-air batteries act as energy stores to buffer intermittent energy supplies from wind or solar plants [15].

6.2.5 Applications in Other Fields

1. In the field of drones—In recent years, with the development of technology, the use of civilian drones has gradually become popular. However, whether it is an aerial photography drone or an operational drone, its weight and range have been bottlenecks limiting its development. To guarantee the long flight time of the drone, it is necessary to provide better energy support for the drone. DJI drones, for example, usually have an outdoor flight time of less than half an hour, and they need to have more components in the battery to ensure long working safety and charge and discharge protection, which undoubtedly increases the cost of the battery. Zinc-air batteries can generate up to 0.3 kWh/kg, which means that although they are half the weight of lithium batteries, zinc batteries have twice the range of lithium batteries. This is a revolutionary increase in weight and range for UAVs, as well as a significant reduction in the cost of use.

2. In the field of special industries—our country is vast and various ecological environments can be found in our country, attracting more researchers and explorers for maintenance and exploration, but once in the no-man's land means losing a lot of support, so it is necessary to carry enough energy supplies at the beginning, such as field exploration operations and nomadic herding For example, in the field of exploration and nomadic herding, where the area of activity is not fixed and there is no fixed place to live, the difficulty of using electricity is an urgent problem, whether it is for the use of tools or for living needs. The high energy and stable power output of zinc air batteries can support the use of field operation equipment and meet the electricity consumption of life.

3. Application in the military field—In the military field, with the massive use of information technology equipment, the system's demand for electrical energy is increasingly high, especially in field combat conditions, electrical energy to supplement the equipment is very little, most of the equipment or the use of diesel engines to supply energy, noise, high heat, easy to be reconnoitered by the enemy, greatly reducing the concealment. At the same time, many night vision goggles, optical targeting equipment and other equipment equipped in a single soldier, the need for miniaturization, high ratio energy electrical energy is more urgent. In addition, the physical requirements of troops and soldiers in military operations are particularly stringent. A US soldier must carry 29.5 kg of supplies, and although a battery weighs only 1 kg, he will have to carry at least six batteries for a three-day mission, taking up 1/5 of his carrying weight. Zinc air batteries are light, safe, and stable, and are not only easy to transport, but can also reduce the burden on the individual soldier, effectively reducing the waste of physical energy and resources [17].

6.3 Summary

In this chapter, two parts have been introduced: the development background of zinc-air batteries and the application sites of zinc-air batteries. The first part mainly includes the advantages of zinc-air batteries, as well as the advantages and disadvantages of other metal-air batteries, the types of zinc-air batteries, the development and research in China. The second part covers the different applications of zinc-air batteries according to their type, mainly button batteries in hearing aids, as a power source in new energy vehicles, as flexible batteries in various wearable devices, and as energy storage devices in the face of wind or solar power plants. Zinc-air batteries, whether in performance or practical use, have the characteristics of high specific energy, large internal energy, smooth discharge curve. It is an energy-saving, environmentally friendly a new type of green battery, not only in the field of new energy electric vehicles or portable electronic products, have a great potential for development. To enhance the application of zinc-air batteries better and faster, the design and preparation of new catalyst materials remain the focus of research.

References

1. Y. Xiao, B. Peng, C. Hu, (Xihua University Industry Press, 2021), p. 280
2. G. Yang, Y. Yang, Q. Wang, H. Wang, (Beijing Chemical Industry Press, 2017), p. 300
3. X. Zou, J. He, L. Guo, H. Yang, C. Wang, C. Li, vol. 48. (Tongren University Industry Press, 2019), p. 66
4. L. Fan, H. Lu, J. Leng, Z. Sun, C. Chen, J. Power Sources **299**, 66 (2015)
5. X. Wang, Z. Li, Y. Qu, T. Yuan, W. Wang, Y. Wu, Y. Li, Chem **5**, 1486 (2019)
6. X. Wang, Z. Chen, S. Chen, H. Wang, M. Huang, Chem. A Eur. J. **26**, 12589 (2020)

7. S. Li, J. Liu, Z. Yin, P. Ren, L. Lin, Y. Gong, C. Yang, X. Zheng, R. Cao, S. Yao, Y. Deng, X. Liu, L. Gu, W. Zhou, J. Zhu, X. Wen, B. Xu, D. Ma, ACS Catal. **10**, 907 (2020)
8. P. Gu, M. Zheng, Q. Zhao, X. Xiao, H. Xue, H. Pang, J. Mater. Chem. A **5**, 7651 (2017)
9. N. Batool, N. Ahmad, J. Liu, X. Han, T. Zhang, W. Wang, R. Yang, J. Tian Mater. Chem. Front. **5**, 2950 (2021)
10. B. Zhu, C. Meng, Y. Guan, S. Guo, J. Hebei Univ. Technol. **49**, 4 (2020)
11. Z. Pei, Z. Yuan, C. Wang, S. Zhao, J. Fei, L. Wei, J. Chen, C. Wang, R. Qi, Z. Liu, Y. Chen, Angew. Chem. Int. Ed. **59**, 4793 (2020)
12. N. Yang, G. Wu, X. Li, vol. 1. (Tianjin Textile Technology, 2022), p. 61
13. S. Li, X. Yang, S. Yang, Q. Gao, S. Zhang, X. Yu, Y. Fang, S. Yang, X. Cai, J. Mater. Chem. A **8**, 5601 (2020)
14. P. Tan, B. Chen, H. Xu, H. Zhang, W. Cai, M. Ni, M. Liu, Z. Shao, Energy Environ. Sci. **10**, 2056 (2017)
15. P. Strasser, Acc. Chem. Res. **49**, 2658 (2016)
16. Z. Tie, Y. Zhang, J. Zhu, S. Bi, Z. Niu, J. Am. Chem. Soc. **144**, 10301 (2022)
17. W. Sun, V. Küpers, F. Wang, P. Bieker, M. Winter, Angew. Chem. Int. Ed. e202207353 (2022)

Chapter 7
Challenges and Prospects for Zinc-Air Batteries

Due to the urgent market demand for green battery products and new energy technologies, a lot of research work have been carried out at home and abroad and significant technological progress has been made, among which electrochemical rechargeable zinc-air secondary batteries with high energy density, safety and environmental protection are gaining more and more attention. Compared with existing chemical power sources, such as lithium-ion batteries, lead-acid batteries, hydrogen–oxygen fuel cells, and lithium-air batteries, and other active metal batteries, zinc-air batteries use a water-soluble electrolyte, with a wide range of raw materials and obvious cost advantages. Zinc-air batteries are superior in the following aspects.

1. High safety: The aqueous solution is used as the supporting electrolyte near room temperature, and the air circulation process can carry heat away from the battery stack, thus completely avoiding the possibility of "thermal runaway" in lithium batteries, which could lead to the combustion of organic solvent electrolytes such as carbonate. Zinc metal is non-toxic and non-hazardous and has the lowest environmental impact throughout the life cycle of the battery.

2. High specific energy: As the positive electrode of the battery uses oxygen from the air as the active material, the capacity is unlimited; the specific energy of the battery depends on the capacity of the negative electrode. A typical zinc-air primary battery has a theoretical specific energy of 1084 Wh kg^{-1}, which is five to six times higher than that of existing lithium-ion batteries. Zinc-air batteries, whether as power batteries for pure electric vehicles or other mobile vehicles, or for energy storage in the process of new energy generation, have a broad development prospect and are the focus of development at home and abroad as the next generation of electrical energy conversion and energy storage technology.

3. Low cost: Battery cost is mainly determined by zinc electrode, air electrode, electrolyte solution and other key components of the battery. In the alkaline electrolyte aqueous solution, it can avoid the use of precious metal catalysts for the preparation of air electrodes. Due to the use of zinc and oxygen in the air as the working medium, the cost of zinc-air batteries is much lower than that of existing

S. Peng, *Zinc-Air Batteries*, https://doi.org/10.1007/978-981-19-8214-9_7

chemical power sources such as lithium-ion batteries and hydrogen–oxygen fuel cells, and is expected to become the preferred technology for future electric vehicle power sources and high-capacity energy storage. However, there is still a long way to the real commercialization and popularity. Here we present some of the challenges and prospects for zinc-air batteries, which focus on improved methods for positive and negative electrode materials and electrolytes and new insights into the development of new integrated zinc-air batteries.

7.1 Problems and Developments in Cathode Materials

The air electrode, as a key component of the battery, is the focus of research and development. Improving the slow kinetic properties of the air electrode in zinc-air batteries remains one of the challenges. Significant developments have been made to improve the performance of zinc-air batteries by modulating the size, morphology, and structure of non-precious metal cathode catalysts. In the development of non-precious metal oxygen reduction catalysts for zinc-air batteries, composite catalysts with high intrinsic catalytic activity and large specific surface area of non-precious metal porous substrate materials compounded with other materials have higher catalytic activity compared to other catalysts and are expected to replace precious metal catalysts. However, there are still some issues to be solved, such as the mechanism of oxygen reduction reaction is not clear, and the cost-effectiveness and scalability of catalytic materials are usually unclear. To rationalize the design of cathode catalysts with bifunctional ORR/OER catalytic activity, the following additional areas of research should be focused on.

1. Development of highly active, stable, and reasonably priced bifunctional electrocatalysts. Further understanding of the mechanism of oxygen reduction and oxygen precipitation processes under aqueous electrolyte conditions, a better understanding of the active sites of catalysts, a clearer understanding of the reactive sites of catalysts and a theoretical basis for catalyst performance analysis will enable the rational design of novel catalysts.
2. Design of structural air electrodes and development of preparation processes. The current research of zinc-air battery is very similar to alkaline fuel cells in that carbon is used as the substrate, but carbon materials are prone to carbon erosion during repeated charging and discharging. Furthermore, the preparation of air electrodes is very tedious, and the addition of auxiliary components can have some detrimental effects. The catalyst is easily dislodged when the force between the catalyst and the substrate is weak, resulting in the degradation of the battery performance. Therefore, there is a need to develop new air electrode structures to make the battery life longer. By improving the air electrode structure, the catalyst can give full play to its catalytic effect.
3. Rational design and development of porous substrate materials with high intrinsic catalytic activity to provide more active sites for the reaction and improve the catalytic performance by compounding with other conductive materials.

4. Development of three-electrode structured zinc-air batteries suitable for batch manufacturing. The decoupling of the ORR and OER processes provides the conditions for maintaining the catalyst and air electrode performance, effectively improving charge and discharge cycle stability. In addition, electrochemically rechargeable zinc-air batteries have obvious advantages in terms of energy density and techno-economics and are an important development direction for new battery technology in the future.

The air electrode design in many zinc-air batteries at the current research stage is similar to common alkaline fuel cells, where the catalyst is mainly supported by a carbon-based gas diffusion electrode. However, the carbon material is susceptible to corrosion (oxidation) under the harsh conditions of repeated discharge and charging in alkaline electrolytes. In addition, the preparation of air electrodes often has inherent disadvantages due to the use of auxiliary materials. Researchers have proposed new strategies highly using corrosion-resistant carbon materials or metal mesh/foam materials. This approach does not require the traditional complex preparation process, which would be of great benefit for large-scale fabrication.

7.2 Problems and Developments in Anode Materials

Zinc-air batteries have good prospects for application not only in electric vehicles, but also in large-scale energy storage and portable electronics. To speed up the marketization of zinc-air batteries, the problem of zinc negative electrode also needs to be solved urgently. In theory, the voltage between the two electrodes of a zinc-air battery is 1.65 V (vs. SHE), however, in practice the discharge voltage is generally lower than 1.2 V and the charging voltage reaches more than 2 V. This directly leads to a Coulomb efficiency of less than 60% for secondary zinc-air batteries, which is mainly caused by dendrite growth, deformation, passivation and hydrogen precipitation corrosion [1]. Currently, the anode material used in commercial primary zinc-air batteries is a paste made of zinc powder, which ensures adequate contact between the active material and the electrolyte, but its rechargeability is extremely poor. Therefore, this type of zinc anode is not suitable for use in secondary zinc-air batteries and researchers have used various strategies to improve zinc anodes. These methods include additive modification, alloy optimization, cladding treatment, zinc complexes and hybrid cells. In addition to modifying the performance of zinc anodes in terms of electrode additives, electrolyte additives, diaphragms and charging and discharging methods, future efforts are needed in the following directions.

1. The discharge products of zinc electrodes exist in the form of zincates. The discharge products of zinc electrodes exist in the form of zincates. Due to the large solubility of zincate in alkaline solutions, the active substance in the zinc electrode is redistributed after several charge and discharge cycles and is unevenly distributed. In some locations the active material is gradually reduced or even completely depleted, in others the active material gradually accumulates and

the electrode becomes thicker, which leads to deformation of the electrode. The deformation of the zinc electrode leads to reduction in the effective area of the electrode, and ultimately low battery life. The study of the mechanism of deformation and passivation of zinc electrodes under alkaline conditions should be strengthened, especially considering that the physical form and structure of the zinc electrode may be different while the dominant mechanism may be different.

2. In practice, $Zn(OH)_4^{2-}$ near the surface of the zinc anode migrates due to natural convection caused by electroosmotic pressure, gravity and other factors, resulting in an extremely uneven distribution on the surface of the zinc anode, which will be uneven after deposition. In the following cycles, the $Zn(OH)_4^{2-}$ in the electrolyte is preferentially deposited on the convex side of the zinc anode because of the higher concentration of $Zn(OH)_4^{2-}$ close to the protruding surface of the zinc anode. With continuous charging and discharging cycles, the dendrites will grow and extend, eventually shedding dead zinc and possibly even piercing the diaphragm and causing a short circuit. As dendrite growth is an inevitable event, it is important to develop efficient additives to inhibit dendrite growth, but also to address or avoid this problem from the physical aspects such as the structural design of the battery assembly.

3. Electrochemically rechargeable secondary batteries are the direction of development for zinc air batteries, and the stability of battery capacity and power during the battery charge/discharge cycle should be fully considered when researching solutions to the problems that exist with the zinc cathode. Researchers in various countries have taken several modification measures: (1) electrode additives; (2) electrolyte additives; (3) zinc alloys; (4) zinc complexes; and (5) coating treatments. These measures have yielded good results in increasing the electrode conductivity, improving the utilization of active substances, inhibiting anode self-corrosion and promoting a uniform distribution of current on the electrode surface. However, in addition to the above improvement measurements, the preparation process of the electrode is also an important part of the battery design. The level of the process determines the specific surface, density and porosity of the electrode, and these factors seriously affect the electrode performance. Therefore, the main development direction for future zinc anodes for secondary zinc-air batteries focuses on the development of new anode materials and controlled processes that can optimize the anode structure. The integrated use of material modification and preparation processes to maximize the electrochemical performance of zinc anodes is key to the commercialization of secondary zinc-air batteries.

Among the metal-air battery systems, zinc-air batteries, including anode, cathode, and electrolyte materials, have received extensive attention from researchers for their numerous advantages. In terms of negative electrodes, zinc metal is the fourth most abundant and cheaply available in the earth's crust. Compared to metallic materials such as aluminium and magnesium, zinc is more chemically stable in alkaline solutions, has a low equilibrium potential and is safe and environmentally friendly. Analyzed from the perspective of the positive electrode, the use of oxygen from the air as a reactant eliminates the need for a special storage tank system and is inexpensive and environmentally friendly. In summary, the development and research of highly efficient and stable performance zinc-air batteries is of great commercial application. China is rich in zinc, magnesium and other mineral resources, and its reserves are among the highest in the world, including 92 million tons of zinc reserves. The development of metal-air batteries for the new energy industry is of great strategic importance for adjusting the structure of the non-ferrous metal industry, giving full play to China's advantages in zinc and magnesium mineral resources, and especially for the development of a highly efficient, clean and safe new energy electric vehicle industry.

7.3 Problems in Developing Electrolytes

As the ionic conductor between the electrodes, the electrolyte plays a fundamental role in the discharge process and has a major impact on the rechargeability, operating voltage, lifetime, power density and safety of zinc-air batteries. As the "blood" of the battery, the electrolyte controls the basic electrochemistry and provides the ionic conductivity in ERZABs. The electrolyte therefore plays a fundamental role not only in the rechargeability and redox reactions of the battery, but also in determining the performance of the ERZABs. In practice, there are often trade-offs between the different properties, which makes it difficult for the electrolyte to meet all requirements.

Based on pH and electrolyte ion type, there are four main categories: (1) alkaline-water-electrolyte-based ERZABs; (2) near-neutral electrolyte ERZABs; (3) dual electrolyte-based ERZABs with an alkaline electrolyte as the anode solution and an acidic solution as the cathode solution; and (4) IL-based ERZABs, whose reaction mechanism varies with the composition of the electrolyte. (i.e. IL cations and anions). For example, depending on the IL anion, the two-electron redox reaction of zinc can yield different complex anions $(Zn(X)_a^{b-})$, where X is the anion, a is the number of anions coordinated to zinc and b is the charge of the complex. With regard to the current problems faced by the above zinc-air battery electrolytes, researchers should carry out further work in the future in the following areas.

1. In alkaline aqueous solutions, several key challenges are detrimental to battery performance. These include the interaction between electrolyte and electrode, carbonization in the ambient atmosphere, and water loss from the electrolyte during the discharge–charge cycle. For example, the discharge product $Zn(OH)_4^{2-}$ is extremely soluble in alkaline electrolytes. However, during battery charging, $Zn(OH)_4^{2-}$ is difficult to deposit sufficiently at the same location on the electrode surface, thus triggering dendritic growth or changes in the electrode shape, leading to decrease in battery performance or, more seriously, to a short circuit. Furthermore, according to the Pourbaix diagram, zinc is thermodynamically unstable in aqueous solutions over the entire pH range. As a result,

zinc tends to dissolve according to the following reactions, which include the undesired production of hydrogen gas:

$$Zn + 2H_2O \rightarrow Zn(OH)_2 + H_2$$

This reaction reduces the efficient use of zinc and can lead to safety problems. In addition, when supersaturated in alkaline aqueous solutions, the discharge products (e.g. $Zn(OH)_4^{2-}$) decompose into insoluble ZnO. The deposition of zinc oxide on the electrode surface leads to passivation of the zinc and reduction in the active area. For air electrodes, carbon-based materials are often used as carriers or catalysts due to their high electrical conductivity and surface area. However, carbon-based electrodes suffer from corrosion problems in alkaline electrolytes, especially when exposed to charge processes. This means that high oxidation potentials can lead to carbon corrosion and impair the long-term function of the electrode. In addition, the semi-open structure of ERZABs leads to challenges with other electrolytes, as follows. One is the commonly used OH-based electrolyte for airborne contaminants (e.g. CO_2), which leads to the formation of carbonate precipitates. These precipitates clog the electrode pores and affect the oxygen reaction, further reducing the operating voltage and energy density of ERZABs. Secondly, the inevitable evaporation of water from the electrolyte can reduce the life of the cell.

2. To alleviate the challenge of the presence of alkaline aqueous solutions, non-alkaline electrolytes with reduced OH^- concentrations have been successfully developed. The use of near-neutral aqueous electrolytes in ERZABs effectively reduces the problems of carbonization and electrode corrosion, which facilitates improved cycle life. However, compared to alkaline electrolytes, near-neutral electrolytes greatly limit the performance of ERZABs due to several key factors such as reaction kinetics and reactant concentrations (e.g. H^+ in near-neutral electrolytes). Firstly, ORR in near-neutral electrolytes exhibits slower kinetics and higher overpotentials compared to alkaline environments, resulting in low discharge voltages and high energy losses in near-neutral electrolyte-based ERZABs. Furthermore, according to the Pourbaix diagram, the air electrode potential is shifted by 59 mV per pH unit. This positive potential shift in the near-neutral electrolyte affects the electric field at the electrolyte–electrode interface, which is detrimental to reducing the binding strength of the adsorbates, especially for charged substances, resulting in slower electrocatalytic reaction kinetics than in alkaline media. Furthermore, the concentration of reactants in the electrolyte (e.g. H^+ in near-neutral electrolytes and OH^- in alkaline electrolytes) has a significant effect on ion conduction and thus on the oxygen reaction process at the air electrode. The concentration of H^+ in a near-neutral electrolyte (pH = 6–7) is much lower than the concentration of OH^- in a basic solution (pH = 12.5–14). Therefore, in near-neutral electrolytes, the reaction process is limited, and the concentration of reactants is significantly lower [2]. The results show that in near-neutral electrolyte-based ERZABs the oxygen reaction reacts slowly and the discharge voltage gap increases.

Carbonation of the alkaline electrolyte leads to: (1) decrease in ionic conductivity, as the mobility of HCO_3^- and CO_3^{2-} is much lower than that of OH^-; (2) the formation of carbonate precipitates, which block the pores in the air electrode; and (3) increase in viscosity, which inhibits the diffusion of oxygen into the electrolyte and leads to sluggish oxygen reaction kinetics. These problems lead to degradation of the capacity and long-term cycling performance of ERZABs. Although attempts have been made to avoid carbonization with CO, the high cost in terms of materials and management limits the applicability of such filters. To address this issue, Schröder et al. investigated the effect of adding K_2CO_3 on alkaline electrolytes in ERZABs and demonstrated that moderate amounts of K_2CO_3 slowed the carbonation reaction, leading to longer cycle life and better practices specific performance. K_2CO_3 was effective in reducing the carbonation kinetics of alkaline electrolytes, thereby extending the service life of ERZABs. However, the increase in cycle stability is at the expense of battery performance as the CO_3^{2-} additive reduces the solubility of oxygen and lowers the ionic conductivity, thereby increasing the internal resistance of the battery and resulting in lower capacity. It is therefore necessary to optimize the composition and concentration of the additives to improve the performance of the electrolyte and the good interaction with the electrodes. In addition to the effects of carbon monoxide, the half-open system of ERZABs leads to unavoidable electrolyte evaporation, which affects the ionic conductivity and viscosity of the electrolyte and thus reduces cell performance. Despite these advances in aqueous solutions, the development of ERZABs still faces several key challenges. Therefore, modification or replacement of alkaline aqueous electrolytes has become a focus of interest to improve the performance of ERZABs.

3. Unlike the electrolyte systems mentioned above, ionic liquids are composed entirely of ions and have negligible vapour pressure and a wide electrochemical window. As a result, ionic liquids have great potential to overcome problems of water loss, carbonization and electrode corrosion. However, it is worth noting that the viscosity of RTILs is several orders of magnitude higher than that of conventional solvents. Ionic viscosity is strongly influenced by the nature of the constituent ions, including their composition, size, molar mass and the interactions between cations and anions. The use of ionic liquids as electrolytes in ERZABs is still in its early stages. Compared to conventional alkaline aqueous electrolytes, in which OH^- is involved in the reaction between zinc and air electrodes, the basic electrochemical properties of ionic liquid-based ERZABs are very different and highly dependent on the nature of the constituent cations and anions. At zinc electrodes, for example, the electrochemistry of zinc is mainly influenced by the type of IL anion, which has a significant impact on the reaction mechanism, redox reversibility, and reaction kinetics. Despite the attractive properties of ionic liquid-based electrolytes for ERZABs, their relatively low conductivity (0.1–40 mS cm^{-1}) is a major limitation. This low conductivity, caused by the low ion mobility due to high viscosity, leads to an inherently sluggish electrochemical kinetics of the redox reactions of zinc and oxygen in ionic liquid based ERZABs.

4. Similar to alkaline aqueous electrolytes, challenges faced by alkaline semi-solid electrolytes include electrode corrosion, zinc passivation, electrolyte water loss and carbonation in the ambient atmosphere. However, unlike aqueous electrolytes, alkaline semi-solid electrolytes offer high mechanical flexibility. In addition, although the use of semi-solid electrolytes does not prevent dendrite formation, the electrolyte can act as a diaphragm to prevent dendrites from causing short circuits in the cell. Some near-neutral chlorine solutions have also been developed to improve self-discharge and eliminate electrolyte carbonation problems that occur in alkaline electrolytes.

5. The development of an effective electrolyte is critical to achieving a breakthrough in ERZAB performance. To this end, researchers need to do a lot of work in elucidating the reaction mechanism of ERZABs and overcoming the shortcomings of specific electrolytes, focusing on (1) the development of the high-performance electrolytes (e.g. near-neutral electrolytes, acidic electrolytes) and specific additives to address key issues such as electrode corrosion, carbonation and water loss; (2) the exploration of the reaction mechanism of different electrolytes on ERZABs. (3) The establishment of a fundamental understanding of the influence of electrolytes on ERZAB performance. Before these efforts can be effectively transferred to ERZABs, the fundamental nature of the electrolyte should be understood. In summary, the degradation of the performance of both liquid-electrolyte based and semi-solid electrolyte based ERZABs can generally be attributed to the following characteristics: failure of the electrolyte in the atmosphere and limited reversibility of the electrode material due to electrolyte–electrode interactions such as corrosion, dendrite growth and passivation. Almost all electrolyte design strategies have been dedicated to solving at least one of these problems. To date, most research has focused on alkaline electrolytes, while non-alkaline and semi-solid electrolytes are still in their early stages. Therefore, a comparison of different types of electrolytes is important to enhance the basic understanding of the fundamentals of electrochemistry and to facilitate further exploration.

6. Theoretical and experimental studies have provided limited fundamental understanding of ionic conduction mechanisms, electrolyte structure–property relationships, zinc dissolution mechanisms of specific electrolytes, electrolyte degradation mechanisms, and failure modes. This requires both theoretical modeling (at the molecular/electronic level) and experimental studies, particularly those involving in situ or on-line characterization such as Fourier transform infrared spectroscopy, nuclear magnetic resonance, Raman microscopy and differential electrochemical mass spectrometry. This fundamental knowledge will elucidate potential detailed reaction mechanisms and guide the development of innovative electrolytes with the right pH and suitable additives, thereby alleviating electrode problems and electrolyte degradation and facilitating the long-term cycling of ERZABs. In addition, the design of electrolytes with good electrode compatibility will likely benefit from the development of machine learning to facilitate the exploration process.

7. As the field of ERZABs becomes increasingly investigated, the development of consistent and clear rules is essential to compare and evaluate the performance of different electrolytes and corresponding ERZABs. For example, the reported current density, energy density and power density of ERZABs are calculated based on different criteria (e.g. area tested, the mass of zinc electrode or active material consumed, volume of the assembled cell). Furthermore, in previous literature, different test durations were used for each cycle when assessing the performance of the charge and discharge cycles. Knowledge of the depth of discharge as well as the mass and volume of the electrolyte is essential to improve the performance evaluation of ERZABs. Furthermore, in order to compare the mechanical properties of flexible semi-solid electrolyte based ERZABs under applied stress, it is necessary to establish standard specifications and methods. The establishment of evaluation criteria that reflect practical applications (e.g. outdoor conditions or extreme temperatures) is also necessary.

7.4 Development of New Integrated Zinc-Air Batteries

Solar cells can only convert light into immediately usable electrical energy and cannot store energy by their nature. Rechargeable zinc-air batteries can store (and release) energy repeatedly as needed through the oxygen precipitation reaction (OER), but this process is inherently plagued by excessive overpotentials caused by sluggish OER kinetics. The combination of these two battery systems would therefore theoretically guarantee the functional integrity of cleanliness and energy efficiency, but research in this area has so far been weak.

The basic structure and working mechanism of the sunlight-promoted zinc-air battery are as follows. It consists of a zinc electrode and a semiconductor photo-electrode, or (α-Fe_2O_3) as an air electrode assembled with an alkaline electrolyte (Fig. 7.1). The discharge process is similar to that of a conventional zinc-air cell: the electrochemical oxidation of Zn at the zinc electrode generates Zn^{2+}, which is accompanied by the reduction of oxygen at the air electrode, producing a current output.

Compared to the OER process in conventional Zn-air batteries, the charging process in light-facilitated rechargeable Zn-air batteries is different in that the charging process under light conditions is significantly easier due to the formation of photogenerated holes with a strong oxidation capacity, which facilitates the removal of protons. During the charging process under solar illumination, the photoelectric electrode absorbs photons from the light source, producing electron–hole pairs. Photogenerated electrons are then rapidly injected into the conduction band (CB) of the semiconductor photoelectrode and further transferred to the Zn electrode via an external circuit, leading to $Zn(OH)_4^{2-}$ to Zn and OH^-. The photoexcited holes simultaneously migrate to the surface of the photoelectrode and oxidize water to oxygen.

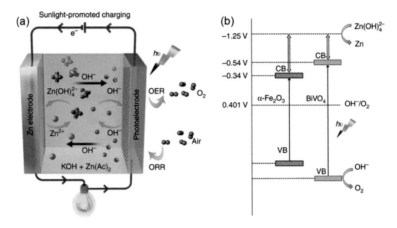

Fig. 7.1 a Basic structural scheme and operating principle of a sunlight-promoted rechargeable zinc-air battery. **b** Mechanism of the sunlight-promoted charging process under solar illumination [3]

This mechanism is expected to lower the charging potential of the Zn-air battery by photo-oxidation of photo-generated holes and facilitate the sunlight-promoted OER reaction during charging. The requirements for the photoelectrode for OH^- to O_2 conversion in sunlight facilitated rechargeable zinc-air batteries should satisfy a valence band (VB) potential above the O_2/OH^- coupling potential [0.401 V relative to the standard hydrogen electrode (SHE)]. When the photoelectrode experiences photoexcitation, the photovoltage is generated to compensate for the high charge potential of the zinc-air battery. It is estimated that sunlight theoretically promotes charge potentials such as the $Zn(OH)_4^{2-}/Zn$ redox potential and quasi-Fermi energy level (EF) for electrons, in its most negative case, close to the conduction band minimum (CBM), at the semiconductor photoelectrode. In summary, the band gap structure of the photoelectrode (e.g. band edge position, band gap) significantly affects the charging performance of sunlight-promoted zinc-air cells.

In the last few years, people have successfully introduced rechargeable zinc-air batteries, whose main advantages are low cost and low environmental impact. Zinc-air batteries are also excellent candidates to replace lead-acid batteries that pose a risk to the environment and human health. With significant advances in the research areas reviewed in this paper, smaller, lighter, and ultimately electrically rechargeable zinc air batteries could also be achieved. Success in these areas could make zinc-air batteries suitable for a wider range of energy storage applications, the most pressing of which is electric vehicles. To date, the relatively high cost and limited energy density of lithium-ion batteries have led to the adoption of electric vehicles relying heavily on government subsidies. This problem can be alleviated by implementing zinc-air batteries, not only because of their lower cost and higher energy density, but also because of their inherent safety, which allows them to use less protective equipment. In addition, high performance and fast discharge capability of the battery is needed for energy bursts, such as vehicle acceleration. In this respect, companies such as Tesla and Mazda have also proposed dual storage systems for electric vehicles, i.e. combining acceleration, regenerative braking and high energy density power sources [4]. The flat discharge curve of zinc-air batteries suggests that this is well suited as a high-energy component for these systems, as they can provide a stable power output in almost any state of charge. The researchers also demonstrated a new hybrid air electrode with a nickel-zinc and air reaction on a rechargeable zinc-air battery, providing very high density and fast discharge capability for electric vehicle propulsion. Meanwhile, the development of miniature and adaptable shape electronics is currently limited by the rigidity and thick casing of current battery technology. Electrically rechargeable zinc-air batteries with flexible solid-state electrolytes are increasingly seen as a viable solution for these applications. While their high volumetric energy density is critical, their safe operation also reduces the need for thick casings, thus further reducing their weight and space requirements.

Finally, zinc-air batteries have received much attention in the academic literature on metal-air chemistry to date, and zinc-air batteries have emerged as the most likely leading energy storage solution. The existing zinc electrode manufacturing infrastructure for alkaline zinc-based batteries and the air electrode design for primary metal air batteries can be used to rapidly scale up the production and commercialization of rechargeable zinc air batteries. Overall, the promising combination of high energy density, safety and low cost should enable electrically rechargeable zinc air batteries to support the energy needs of an increasingly affluent, digital, and low carbon global economy. We therefore highly encourage accelerated research and development of this technology in both the academic and industrial sectors.

References

1. J. Fu, Z.P. Cano, M.G. Park, A. Yu, M. Fowler, Z. Chen, Adv. Mater. **29**, 1604685 (2017)
2. J. Fu, R. Liang, G. Liu, A. Yu, Z. Bai, L. Yang, Z. Chen, Adv. Mater. **31**, 1805230 (2019)
3. X. Liu, Y. Yuan, J. Liu, B. Liu, X. Chen, J. Ding, X. Han, Y. Deng, C. Zhong, W. Hu, Nat. Commun. **10**, 4767 (2019)
4. J. Song, C. Wei, Z. Huang, C. Liu, L. Zeng, X. Wang, Z.J. Xu, Chem. Soc. Rev. **49**, 2196 (2020)

Printed in the United States
by Baker & Taylor Publisher Services